Thermomechanical Simulation Methodologies for Advanced Semiconductor Packaging

Other related titles:

You may also like

- PBCS028 | Ogrenci-Memik | Heat Management in Integrated Circuits | pub November 2015
- PBCS073 | Dhiman | VLSI and Post-CMOS Electronics, Volume 1: Design, modelling and simulation | pub September 2019
- PBCS073 | Dhiman | VLSI and Post-CMOS Electronics, Volume 2: Devices, circuits and interconnects | pub September 2019

We also publish a wide range of books on the following topics:
Computing and Networks
Control, Robotics and Sensors
Electrical Regulations
Electromagnetics and Radar
Energy Engineering
Healthcare Technologies
History and Management of Technology
IET Codes and Guidance
Materials, Circuits and Devices
Model Forms
Nanomaterials and Nanotechnologies
Optics, Photonics and Lasers
Production, Design and Manufacturing
Security
Telecommunications
Transportation

All books are available in print via https://shop.theiet.org or as eBooks via our Digital Library https://digital-library.theiet.org.

IET MATERIALS, CIRCUITS & DEVICES SERIES 85

Thermomechanical Simulation Methodologies for Advanced Semiconductor Packaging

Shuye Zhang and Guoli Sun

The Institution of Engineering and Technology

About the IET

This book is published by the Institution of Engineering and Technology (The IET).

We inspire, inform and influence the global engineering community to engineer a better world. As a diverse home across engineering and technology, we share knowledge that helps make better sense of the world, to accelerate innovation and solve the global challenges that matter.

The IET is a not-for-profit organisation. The surplus we make from our books is used to support activities and products for the engineering community and promote the positive role of science, engineering and technology in the world. This includes education resources and outreach, scholarships and awards, events and courses, publications, professional development and mentoring, and advocacy to governments.

To discover more about the IET please visit https://www.theiet.org/.

About IET books

The IET publishes books across many engineering and technology disciplines. Our authors and editors offer fresh perspectives from universities and industry. Within our subject areas, we have several book series steered by editorial boards made up of leading subject experts.

We peer review each book at the proposal stage to ensure the quality and relevance of our publications.

Get involved

If you are interested in becoming an author, editor, series advisor, or peer reviewer please visit https://www.theiet.org/publishing/publishing-with-iet-books/ or contact author_support@theiet.org.

Discovering our electronic content

All of our books are available online via the IET's Digital Library. Our Digital Library is the home of technical documents, eBooks, conference publications, real-life case studies and journal articles. To find out more, please visit https://digital-library.theiet.org.

In collaboration with the United Nations and the International Publishers Association, the IET is a Signatory member of the SDG Publishers Compact. The Compact aims to accelerate progress to achieve the Sustainable Development Goals (SDGs) by 2030. Signatories aspire to develop sustainable practices and act as champions of the SDGs during the Decade of Action (2020–30), publishing books and journals that will help inform, develop, and inspire action in that direction.

In line with our sustainable goals, our UK printing partner has FSC accreditation, which is reducing our environmental impact to the planet. We use a print-on-demand model to further reduce our carbon footprint.

British Library Cataloguing in Publication Data
A catalogue record for this product is available from the British Library

ISBN 978-1-83724-140-8 (hardback)
ISBN 978-1-83724-141-5 (PDF)

Typeset in India by MPS Limited

Cover image: Yuichiro Chino/Moment via Getty Images

Contents

Preface

In the field of advanced electronic packaging today, numerical simulation technology plays an essential role. As electronic devices evolve toward miniaturization, thinness, high performance, and multifunctionality, packaging technologies continue to advance, setting higher standards for the reliability and performance of electronic packaging. Numerical simulation technology, particularly finite element analysis (FEA), has become vital for researching and optimizing electronic packaging structures. This technology enables precise simulation and analysis of multi-physics issues in electronic packaging, including thermal management, mechanical behavior, and electromagnetic compatibility. Through simulation, engineers can predict and address potential problems in the early stages of product development, reducing costs, shortening time-to-market, and enhancing product reliability. For instance, in TSV technology, numerical simulation is used to study the growth mechanism of copper during electroplating, optimize process parameters, and reduce experimental costs.

Additionally, FEA plays a significant role in evaluating the thermal fatigue life of solder joints, controlling and optimizing package warpage, and studying thermal performance. Numerical simulation technology also promotes the development of experiments by providing more reliable theoretical guidance for the scientific formulation of test plans, the determination of the best measurement points during experiments, and the setting of instrument ranges. It can intuitively display phenomena that are not readily observable and difficult to explain, facilitating understanding and analysis. It can also reveal internal physical phenomena that no experiment can see. Moreover, numerical simulation can replace dangerous, expensive, or even difficult-to-implement experiments, such as reactor explosion accidents and the processes and effects of nuclear explosions. With the advancement of computer technology, the application of numerical simulation technology in electronic packaging is becoming more widespread. It provides an in-depth understanding of complex physical phenomena and accelerates the development and innovation of new products. Therefore, numerical simulation technology is indispensable in driving the growth of electronic packaging technology and improving product performance and reliability.

This book primarily focuses on studying electronic packaging in assembly manufacturing processes and the failure mechanisms in assembly manufacturing and testing through modeling and simulation. The basic framework of the book is divided into eight chapters. Chapter 1 provides an overview of the rapid and ongoing evolution in packaging technology over recent decades. Packaging technology has made significant strides in size, performance, and functionality from early dual in-line packages (DIP) to modern surface mount technology (SMT) and advancing to today's 3D integrated circuits and heterogeneous integration. This section also explores the current status of advanced packaging technologies and future trends. Electronic packaging encompasses diverse disciplines such as thermology, mechanics, and electromagnetism. Various design solutions can be swiftly and accurately evaluated through numerical simulations, reducing development costs, accelerating product launches, and fostering continual innovation in electronic packaging technology. Moreover, this chapter emphasizes the fundamental computational principles of the finite element method and its application in assessing the thermomechanical reliability of electronic packaging structures. Chapter 2 includes the hierarchical multiscale method proposed in this study is utilized to develop a high-precision prediction model for TSV wafer warpage. The established equivalent model accurately forecasts the evolution of wafer warpage throughout all TSV wafer encapsulation processes. Furthermore, the impact of varying the number of interposer partitions on prediction accuracy is systematically investigated, leading to the formulation of an optimization strategy. Chapter 3 proposes a coupled analysis method based on the representative volume element (RVE) technique and the generalized Hill plasticity model. The coupling algorithm extracts plastic parameters from the homogenization region of micro solder joints in 2.5D packaging structures and inputs them into a homogeneous model to replace the detailed model. Chapter 4 provides three-dimensional mold flow analysis software to simulate the transfer molding process, providing theoretical analysis guidance before actual encapsulation processes. Chapter 5 is concerned with the stress of PBGA package solder joints under vibrational loads using the finite element model and performs numerical analyses for both harmonic sweep and random vibration. Chapter 6 is mainly about the simulation results revealing significant current accumulation at the inner corners of the metal interconnect lines, where electromigration is most severe. Chapter 7 describes a modeling method based on fracture mechanics to simulate residual stress distribution in TSV/RDL interconnect structures post-annealing. Besides, the influence of TSV/RDL geometric and thermomechanical parameters on the interfacial cracking behavior of the interconnect structures is also investigated. The final chapter will introduce the dynamic response of solder joints in board-level BGA packages during drop impacts, which was analyzed using the explicit finite element dynamic analysis method.

Acknowledgments

The completion of this book was made possible through considerable support and assistance, for which we are sincerely grateful. We extend our heartfelt thanks to the Institute of Electronics Packaging Technology and Reliability at Beijing University of Technology for their support. We also thank all the professors and students for their academic guidance and assistance. Your valuable opinions and suggestions have played a significant role in the completion of this book.

About the authors

Shuye Zhang is an associate professor at Harbin Institute of Technology, China. He has been working on electronic packaging areas for 15 years. Currently, he is working on heterogeneous integration, 2.5D packaging, and solder joint reliability.

Guoli Sun is currently pursuing a Ph.D. at Harbin Institute of Technology, where his research focuses on material and structural reliability in advanced electronic packaging, as well as the mechanical characterization and performance analysis of various materials.

Chapter 1

Advanced semiconductor electronic packaging

1.1 Introduction

Since the 1950s, human civilization has entered its third technological revolution following the steam and electricity revolutions, marked notably by the advent of electronic information technology. This revolutionary technology, akin to others, has not only greatly propelled transformations in human society encompassing economics, politics, culture, and beyond but has also significantly altered human lifestyles and modes of thinking [1]. Today, the level of advancement in electronic information technology has become a crucial indicator of a country's comprehensive national strength and international competitiveness. The electronic information industry has emerged as a strategically vital and overarching pillar within the national economy. As the foundation and core of electronic information technology, semiconductor integrated circuit (IC) technology has thus evolved into a highly competitive, innovative, and immensely promising development field. With the rapid progress of electronic information technology, three major relatively independent industrial sectors have gradually emerged globally: design, manufacturing, and packaging and testing, serving as indispensable pillars of the industry. Among these, the packaging and testing sector plays a significant role in driving the overall development of the IC industry and also catalyzes advancements in other foundational industries.

Electronic packaging serves as the bridge between chips and external systems, providing essential support and protection for chips while ensuring their functionality. However, in recent years, as semiconductor manufacturing technology approaches the process and physical limits of complementary metal-oxide semiconductor (CMOS), the longstanding "Moore's Law" revered by the industry will no longer be universally applicable [2,3]. Consequently, electronic packaging must assume responsibilities beyond mechanical support, environmental protection, heat dissipation, and electrical interconnection for IC chips. It must also enhance circuit performance, facilitate system integration, rather than solely rely on integrating more functions onto a single chip to improve device performance. This development trend has spurred the emergence of advanced packaging technologies such as system-in-package (SiP), two-dimensional/three-dimensional (2.5D/3D) packaging, chip-on-wafer-on-substrate (CoWoS) technology, and through-silicon via (TSV) interconnects [4–7].

1.2   Development of electronic packaging

The invention of the first transistor in 1947 at Bell Labs marked a revolutionary shift in the landscape of electronics and heralded the beginning of the IC era. This innovation, which replaced the bulky and unreliable vacuum tubes used in early electronics, laid the foundation for modern semiconductor technology. The subsequent advent of ICs in the 1950s and 1960s enabled the miniaturization of electronic devices and the development of complex systems on a single chip, significantly improving performance, power efficiency, and cost-effectiveness. In 1965, Dr. Gordon Moore, co-founder of Intel, introduced Moore's Law, which predicted that the number of transistors that could be integrated into a semiconductor chip would double approximately every 18 months. This observation not only provided a benchmark for technological progress but also became a guiding principle for the semiconductor industry, driving continuous innovation for several decades [8].

With the rapid advancements in electronic information technology, the semiconductor industry has evolved into a multifaceted ecosystem that encompasses three distinct yet highly interdependent sectors: design, manufacturing, and packaging and testing. These sectors have become the core pillars of the IC industry, each contributing to the development of more advanced and efficient technologies. The design sector focuses on creating circuit layouts and architectures that maximize performance while minimizing power consumption and physical space. The manufacturing sector is responsible for fabricating the semiconductor chips, utilizing sophisticated photolithography and deposition techniques to produce smaller, more densely packed transistors. Packaging and testing, often considered the final stage in the semiconductor production process, plays an equally critical role by protecting the delicate circuits and ensuring the reliability and functionality of the chips under real-world conditions. This sector has become increasingly important as the size of semiconductor devices has decreased, and the need for high-density, high-performance packaging solutions has grown [9].

While the packaging and testing sector has long been integral to the semiconductor industry, its role has become even more pivotal in recent years. Advances in packaging technologies—such as SiP, 3D packaging, and flip-chip designs—have enabled the continued scaling of ICs, allowing chips to meet the ever-increasing demands of miniaturization and performance. As semiconductor devices are packed with more transistors and become more complex, the need for robust packaging solutions that ensure thermal management, electrical connectivity, and mechanical integrity has never been more critical. Furthermore, packaging technologies directly impact a device's performance, reliability, and overall cost, making it a key driver of innovation in the broader electronics industry.

However, the rapid pace of progress in semiconductor manufacturing and the physical limitations of current technologies have presented significant challenges. As CMOS technology approaches its theoretical limits in terms of miniaturization and power efficiency, the continuing applicability of Moore's Law is being questioned. The reduction in transistor size, while enabling more powerful devices, also introduces issues such as heat dissipation, signal interference, and leakage currents. As a result, the industry is shifting toward alternative materials and architectures, including quantum computing, neuromorphic computing, and 3D stacking, which promise to extend the potential of ICs beyond the limits of traditional CMOS scaling [10,11].

In response to these challenges, the International Technology Roadmap for Semiconductors (ITRS) has outlined a new developmental pathway for the semiconductor industry. This roadmap, illustrated in Figure 1.1, presents a strategic vision that integrates next-generation semiconductor technologies, such as advanced packaging, heterogeneous integration, and innovative material systems. The ITRS aims to address the limitations of Moore's Law by exploring new technological trajectories that enable continued progress in the face of scaling challenges. These include innovations in photonics, quantum devices, and alternative computing paradigms that could redefine the future of ICs [12].

As the semiconductor industry enters the post-Moore era, device feature sizes are approaching their physical limits, signifying a pivotal shift in technological advancement [13]. However, the rapid advancement of the Internet of Things, smart terminals, and industrial intelligence has generated a pressing demand for high-density integrated, multifunctional, and low-power electronic devices. To better address this burgeoning market need, many companies are shifting their focus from simply reducing chip feature sizes to enhancing electronic packaging technology. Electronic packaging serves as a critical interface connecting chips with external systems, providing essential support and protection while ensuring their functionality. It creates a stable and reliable operational environment for the chip, facilitates input and output connections to the outside world, and efficiently dissipates internal device heat. Through rigorous performance testing,

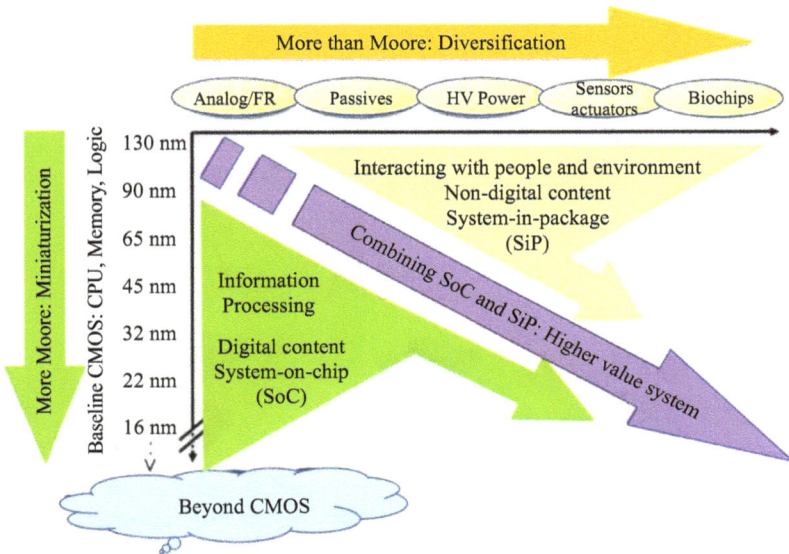

Figure 1.1 The roadmap for semiconductors [12]

screening processes, and various environmental and mechanical experiments, packaging technology guarantees chip reliability, enabling stable and optimal functionality [14].

Electronic packaging technology has evolved alongside the advancement of chips, embodying the principle that "each generation of chips requires a corresponding generation of packaging." The history of packaging technology mirrors the continuous enhancement of chip performance and the ongoing miniaturization of systems. In essence, the development of electronic packaging technology can be delineated into five distinct stages, each characterized by its typical packaging forms, illustrated in Figure 1.2.

Since the 1950s, electronic packaging technology has evolved through several distinct stages, closely intertwined with advancements in chip capabilities. This evolution follows the fundamental principle that each new generation of chips necessitates a corresponding evolution in packaging. The historical development of packaging technology mirrors the relentless improvement in chip performance and the progressive miniaturization of systems. The initial stage introduced the zero-level package in the 1950s, followed by the transition to first-level packaging in the 1960s and the subsequent progression to second-level packaging by the 1970s. The second stage, prior to the 1980s, marked the era of through-hole insertion technology, typified by packages such as the transistor outline and dual-in-line package (DIP). During this period, packaging structures were relatively simplistic in function, featured fewer pins (typically fewer than 64), and had fixed pin pitches. Increasing the number of pins resulted in larger package sizes, limiting the maximum mounting density to approximately 10 pins/cm^2. The advent of surface mount technology in the post-1980s era constituted the third stage, epitomized by packages like the small outline package (SOP) and quad flat package (QFP). This phase saw a significant increase in pin count and mounting density, with some

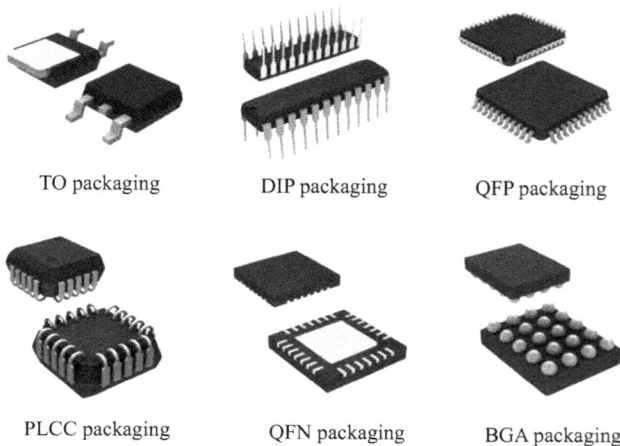

TO packaging   DIP packaging   QFP packaging

PLCC packaging   QFN packaging   BGA packaging

Figure 1.2   Several typical package types

packages accommodating up to 300 pins and achieving mounting densities ranging from 10 to 50 pins/cm^2. Surface mount technology revolutionized packaging by enabling higher pin counts and improved integration capabilities. The fourth stage emerged in the 1990s with the introduction of packages such as the ball grid array (BGA) and chip scale package (CSP). This era encompassed advancements including BGA, CSP, SiP, and wafer-level CSP. BGA packages, featuring pin pitches primarily of 1.5 and 1.27 mm, achieved installation densities ranging from 40 to 60 pins/cm^2. Each of these stages represents a critical juncture in the evolution of electronic packaging technology, characterized by distinctive packaging forms and technological advancements, as depicted in Figure 1.2.

As chip sizes increase, integration intensifies, functionalities strengthen, and I/O capabilities expand, electronic packaging technology faces escalating demands. This evolution is steering away from traditional lead and pin bonding toward the trend of 2.5D/3D packaging, characterized by miniaturization, high density, enhanced performance, superior reliability, and cost-effectiveness. This represents the fifth stage in the development of electronic packaging technology, spanning from the late 1990s to the present era. This advancement underscores the critical need for packaging solutions that can accommodate larger chips while meeting rigorous performance standards, thus driving innovation toward more sophisticated packaging architectures [15]. Figure 1.3 illustrates the major developments in the form of IC packaging over the years. In response to the escalating demand for chips, a variety of advanced packaging technologies have emerged, each aimed at enhancing packaging density and performance while maintaining cost-effectiveness and reliability. Advanced packaging has become a pivotal direction for the development of the IC industry, driving innovations in materials, design, and functionality to support the integration of increasingly complex and sophisticated chip architectures. This evolution is critical to enabling the continued miniaturization and performance enhancement of electronic devices, underscoring the profound impact of packaging technology on the broader semiconductor industry.

In response to the rapid development and evolving demands of the semiconductor industry, leading packaging and testing companies such as Huatian Technology, TSMC, Toshiba, Intel, Samsung, and SPIL have introduced a wide range of advanced packaging technology solutions. These innovations encompass various cutting-edge approaches, including SiP, 3D packaging, wafer-level packaging, panel-level packaging, and embedded multi-chip interconnect bridge (EMIB). Each of these technologies plays a crucial role in enhancing the performance, integration, and scalability of electronic systems. Among these, EMIB and CoWoS technologies have gained widespread application, particularly in high-performance computing domains such as Field-Programmable Gate Arrays, Central Processing Units, Graphics Processing Units, and other specialized chips. These advanced packaging solutions are particularly well-suited for the increasing demands of data centers, artificial intelligence (AI), and machine learning, where high bandwidth, low latency, and high power efficiency are essential.

In recent years, the Chiplet architecture has emerged as a transformative solution, further propelling the adoption of 2.5D and 3D packaging techniques.

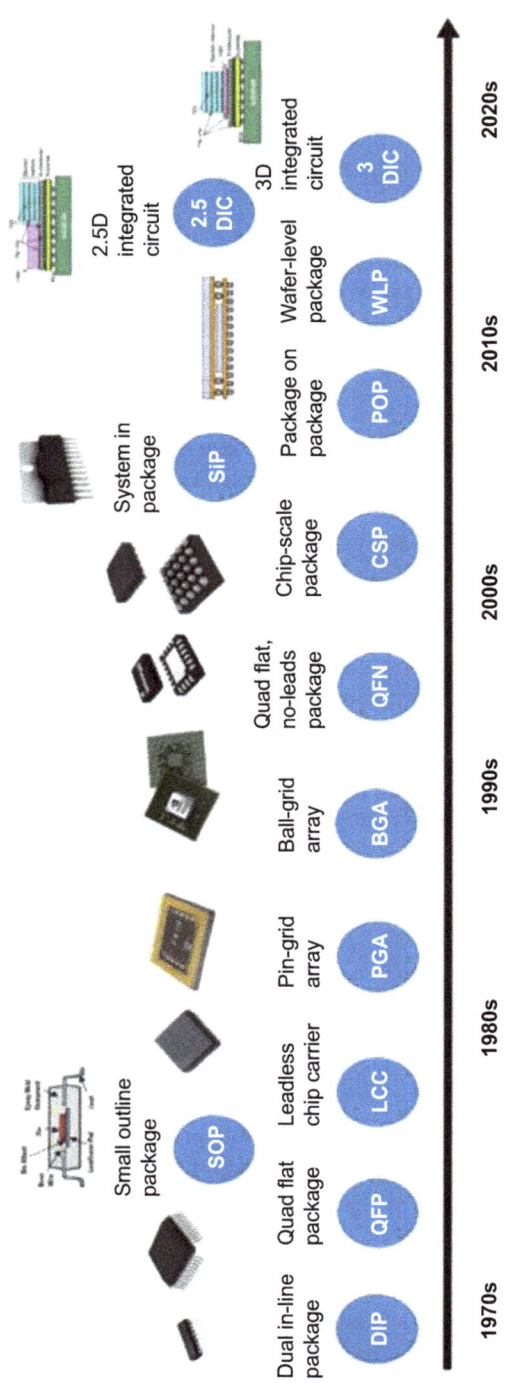

Figure 1.3 The development of IC packaging technology [15]

This innovative approach involves partitioning large, monolithic chips into smaller, independent functional modules, known as chiplets. Each chiplet can be optimized for specific tasks, enabling greater flexibility in chip design and manufacturing. This modular approach allows for the integration of diverse functionalities, leading to a more efficient use of silicon real estate and reducing design and production costs. Moreover, chiplets offer the potential for heterogeneous integration, allowing different process technologies to coexist within a single package, which significantly enhances system performance. The growing prominence of 2.5D and 3D packaging in chiplet-based products marks a new era of semiconductor design, where vertical stacking and multi-layer integration provide higher performance, improved power efficiency, and better thermal management. These technologies allow for more compact form factors without compromising functionality or performance. The continuous evolution and widespread adoption of advanced packaging and interconnect technologies are not only driving miniaturization but also enabling the integration of increasingly sophisticated functionalities into smaller, more efficient devices.

As the semiconductor industry pushes the boundaries of performance and functionality, advanced packaging technologies will play an even more critical role in shaping the future landscape of electronic devices and computing systems. By facilitating the seamless integration of diverse components, these technologies will be at the heart of innovation, enabling next-generation computing, communications, and consumer electronics. Ultimately, advanced packaging is key to meeting the growing demand for more powerful, compact, and energy-efficient devices that are capable of supporting emerging applications such as 5G, AI, and quantum computing.

1.3 Thermomechanical numerical simulation

Encapsulation structures, integral to modern electronics, represent complex systems characterized by multi-materials, multi-interfaces, and multi-scale dimensions. These structures, which are designed to protect delicate semiconductor components, face various challenges during their lifecycle, including processing, manufacturing, testing, and in-service use. Among these challenges, stress concentration and micro-cracking are particularly prevalent. Stress concentration is most significant at the corners and edges of encapsulation structures, where geometric irregularities or changes in material properties amplify internal stresses. These regions often serve as the initiation sites for crack propagation, which can compromise the mechanical integrity of the device. Additionally, the interfaces between different materials in the encapsulation structure are prone to cracking due to their inherent lower strength relative to the bulk material, further exacerbating reliability concerns. Traditional experimental methods for evaluating these issues are often time-consuming, inefficient, and costly, making them less suitable for the rapid optimization cycles required in advanced electronic packaging design.

To overcome these challenges, finite-element analysis (FEA) has emerged as an essential computational tool in the field of electronic packaging. FEA offers a versatile approach to simulate and analyze complex mechanical behaviors within encapsulation structures. By providing insights into stress, strain, creep fatigue, crack initiation, and propagation, FEA enables engineers to better understand the performance of packaging materials and structures under different loading and environmental conditions. In particular, FEA has proven to be invaluable for optimizing material selection, refining structural designs, and performing reliability analyzes to ensure the long-term performance and durability of electronic devices. As a result, FEA has become an indispensable part of the design and analysis workflow in electronic packaging, significantly enhancing the speed and accuracy of development while reducing the costs associated with traditional experimental techniques [16–18].

The finite-element method (FEM) was first introduced by Clough in 1960 [19], offering a systematic way to simulate and analyze the mechanical behavior of complex structures by subdividing them into simpler, manageable subdomains called finite elements. Each of these elements is interconnected at nodes, which are points where the elements meet. The overall behavior of the structure is determined by solving a system of equations that describe the relationships between the nodal forces and displacements, derived from the principles of mechanics and the variational method. The discretization process used in FEM allows engineers to approximate the continuous behavior of a material or structure, making it possible to solve complex engineering problems that would otherwise be intractable.

The development and optimization of finite element software have played a pivotal role in advancing the applicability of FEM. Researchers and engineers can now accurately model and predict the behavior of structures under various loading conditions, providing crucial insights into design optimization. FEM has expanded beyond traditional structural analysis and is now capable of handling multi-physics coupling, such as thermal, electrical, and mechanical interactions. This extension allows for a more comprehensive analysis, addressing not only mechanical stresses but also issues related to thermal effects, electrical conductivity, and material behaviors under extreme conditions. This multi-disciplinary approach is particularly beneficial in the field of electronic packaging, where devices are often subject to coupled thermal-mechanical loading conditions, requiring a more integrated approach to design and reliability analysis.

FEM discretizes the structure into finite elements, where each element is connected by nodes. The behavior of each element is described by a set of equations based on the mechanical properties of the material, such as elasticity, plasticity, and viscosity. These equations are then solved using numerical methods to determine the displacements at the nodes. The primary unknowns in FEM are the nodal displacements, which are calculated by solving a system of algebraic equations derived from the element's mechanical characteristics. Interpolation functions are then used to approximate the displacement field over the entire structure. The accuracy of the FEM solution depends on the size and number of elements used—smaller elements with more subdivisions result in a more accurate solution, but at the cost of increased computational complexity. Convergence studies are performed to ensure that the solution is accurate and reliable [20,21].

As FEM theory has matured and computational power has increased, the capabilities of FEA have expanded significantly. Initially focused on linear elastic materials, FEA now includes a wide range of material models, including elastoplastic, viscoelastic, and viscoplastic behaviors, allowing it to simulate real-world material responses more accurately. Moreover, FEA has extended beyond solid mechanics to include the analysis of fluid dynamics, heat transfer, and electromagnetic fields. These capabilities are particularly relevant for electronic packaging, where interactions between mechanical stresses, thermal gradients, and electrical currents are critical to device performance and reliability. Thermo-electro-mechanical coupling is one such multi-physics phenomenon that FEA can model, allowing for a more holistic under-standing of how thermal cycles, electrical fields, and mechanical forces interact within a packaged electronic device.

Commercial FEA software such as ABAQUS, ANSYS, and COMSOL has been developed to support these advanced capabilities, becoming essential tools in the field of electronic packaging. These software packages are widely used in industry for a range of applications, from simulating thermal management and stress dis-tribution within package structures to evaluating the fatigue life of solder joints and predicting failure modes such as interface delamination. For example, FEA tools are used to study the thermo-mechanical reliability of TSVs, investigate the evolution of wafer warpage during packaging, estimate the fatigue life of solder joints, simulate plastic sealing processes, and assess electromigration in interconnects, as shown in Figure 1.4. These simulations help to optimize package designs, identify potential

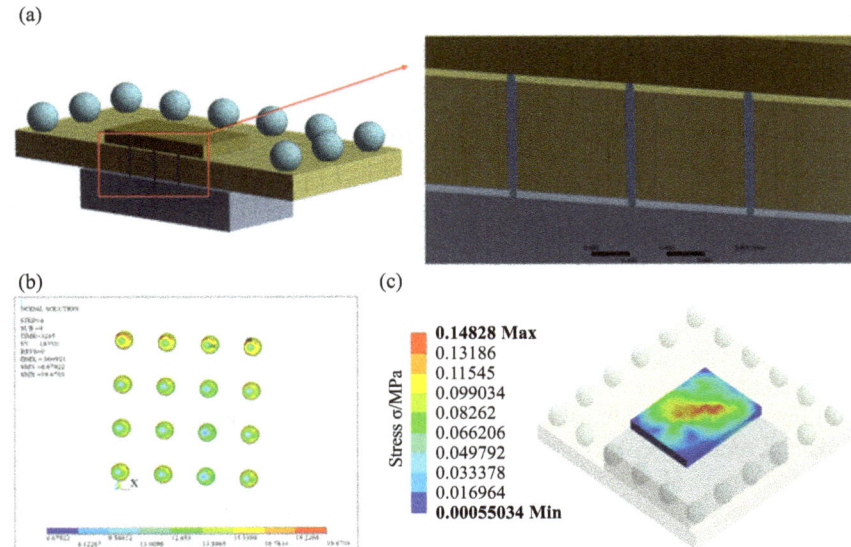

Figure 1.4   *Applications of FE simulation for electronic packaging structures: (a) element model of 3D packaging structure [22], (b) stress contour of solders in the normal direction, and (c) stress contour of die under temperature loading*

failure points early in the design process, and improve the overall performance, reliability, and manufacturability of electronic devices [23–26].

Furthermore, the integration of FEA into the design and manufacturing workflows has led to significant improvements in the efficiency and cost-effectiveness of packaging development. Through virtual prototyping and simulation-driven design, engineers can rapidly test and refine multiple design iterations without the need for extensive physical testing. This capability not only accelerates the development cycle but also enables the creation of more reliable and robust packaging solutions that meet the increasingly stringent demands of modern electronic devices, including smaller form factors, higher performance, and greater power efficiency.

1.4   Chapter summary

This chapter provides an in-depth overview of the rapid evolution of packaging technology within the electronics industry, highlighting key advancements that have significantly enhanced the performance, size, and functionality of electronic devices. Beginning with the early DIP, which were widely used in the 1970s and 1980s, electronic packaging was initially limited to simple designs that involved through-hole mounting of components. As technology progressed, the introduction of surface-mount technology revolutionized the industry, enabling components to be directly mounted onto the surface of the circuit board, which allowed for smaller, more compact devices and improved automation in manufacturing. In recent years, more sophisticated packaging technologies such as three-dimensional integrated circuits (3D ICs) and heterogeneous integration have emerged, representing a significant leap forward in terms of performance and miniaturization.

The chapter explores the current state and future trends in advanced packaging technologies. One prominent trend is the integration of chiplets, which are small, modular chips that can be combined in various configurations to meet the specific performance needs of a system. This modular approach not only offers flexibility but also allows for better performance scaling and cost-effective manufacturing. Additionally, miniaturization remains a driving force, with ongoing efforts to create smaller components without sacrificing power or functionality. Moreover, advances in interconnects, such as TSVs and micro bumps, are improving data transfer speeds and reducing energy consumption in advanced packages.

The chapter also emphasizes the growing importance of interdisciplinary fields, such as thermology, mechanics, and electromagnetism, in the development of electronic packaging solutions. Engineers now rely heavily on numerical simulations to optimize designs before physical prototypes are built. These simulations enable engineers to assess various design configurations, reducing development costs and time to market. Among the most widely used simulation techniques is the FEM, which helps predict potential failure modes and optimize the thermo-mechanical reliability of packaging structures. By simulating stresses such as thermal cycling and mechanical loads, FEM ensures that designs are durable and capable of withstanding real-world conditions.

In conclusion, packaging technology has made impressive strides in recent decades, enabling the creation of smaller, faster, and more reliable electronic devices. With continuous advances in simulation tools, materials, and innovative packaging techniques, the field will continue to drive the development of cutting-edge electronic systems across a wide range of industries, from consumer electronics to high-performance computing and telecommunications. As the demands for higher performance and energy efficiency continue to grow, packaging technologies will remain central to the future of electronic innovation.

References

[1]   J. Fitzsimmons, Information technology and the third industrial revolution, *The Electronic Library* 12 (1994) 295–297.

[2]   M.M. Waldrop, The chips are down for Moore's law, *Nature* 530 (2016) 144–147.

[3]   G. Chirkov and D. Wentzlaff, Seizing the bandwidth scaling of on-package interconnect in a post-Moore's law world, in: *Proceedings of the 37th International Conference on Supercomputing*, ACM, Orlando, FL, 2023: pp. 410–422.

[4]   S. Zhang, Z. Li, H. Zhou, *et al.*, Challenges and recent prospectives of 3D heterogeneous integration, *E-Prime – Advances in Electrical Engineering, Electronics and Energy* 2 (2022) 100052.

[5]   S. Zhou, K. Ma, Y. Wu, *et al.*, Survey of reliability research on 3D packaged memory, *Electronics* 12 (2023) 2709.

[6]   S.Y. Hou, W.C. Chen, C. Hu, *et al.*, Wafer-level integration of an advanced logic-memory system through the second-generation CoWoS technology, *IEEE Transactions on Electron Devices* 64 (2017) 4071–4077.

[7]   J.H. Lau, Recent advances and trends in advanced packaging, *IEEE Transactions on Components, Packaging and Manufacturing Technology* 12 (2022) 228–252.

[8]   G.E. Moore, Cramming more components onto integrated circuits, *Proceedings of the IEEE* 86 (1998) 82–85.

[9]   National Research Council, Division on Engineering and Physical Sciences, Commission on Engineering and Technical Systems, *et al. Microelectromechanical Systems: Advanced Materials and Fabrication Methods*, National Academies Press, Washington, DC, 1998.

[10] S.E. Thompson and S. Parthasarathy, Moore's law: the future of Si microelectronics, *Materials Today* 9 (2006) 20–25.

[11] S. Li, *From Moore's Law to Function Density Law*, Springer Nature, Singapore, 2022: pp. 3–27.

[12] G.Q. Kouchi Zhang, M. Graef, and F. Van Roosmalen, The rationale and paradigm of "More than Moore," in: *56th Electronic Components and Technology Conference 2006*, IEEE, San Diego, CA, 2006: pp. 151–157.

[13]   J.Y. Xu and A.J. Huang, Technological innovation in the post-Moore law period, *Electronics & Packaging* 20 (2020) 3–6.

[14]   Z.W. Chen, Y.H. Mei, S. Liu, *et al.*, Reliability in electronic packaging: past, now and future, *Journal of Mechanical Engineering* 57 (2021) 248–268.

[15]   X.Y. Zhou, Overview of advanced IC packaging technology, *Application of IC* 35 (2018) 1–7.

[16]   J.R. Lee, M.S.A. Aziz, M.H.H. Ishak, and C.Y. Khor, A review on numerical approach of reflow soldering process for copper pillar technology, *The International Journal of Advanced Manufacturing Technology* 121 (2022) 4325–4353.

[17]   H. Cui, W. Tian, H. Xu, *et al.*, The reliability of the complex components under temperature cycling, random vibration, and combined loading for airborne applications, *Crystals* 13 (2023) 473.

[18]   W. Yu, S. Cheng, Z. Li, *et al.*, The application of multi-scale simulation in advanced electronic packaging, *Fundamental Research* 4 (2024) 1442–1454.

[19]   R.W. Clough, The finite element method after twenty-five years: a personal view, *Computers & Structures* 12 (1980) 361–370.

[20]   W.K. Liu, S. Li, and H.S. Park, Eighty years of the finite element method: birth, evolution, and future, *Archives of Computational Methods in Engineering* 30 (2023) 3475–3475.

[21]   Z. Pater, The application of finite element method for analysis of cross-wedge rolling processes—a review, *Materials* 16 (2023) 4518.

[22]   Y. Xia, Y. Su, X. Xu, *et al.*, Effect of thermal cycling on the strain and stress distribution of TSVs and solder bumps in stacked package structure, *E-Prime – Advances in Electrical Engineering, Electronics and Energy* 5 (2023) 100274.

[23]   M. Zhang, F. Chen, F. Qin, S. Chen, and Y. Dai, Correlations between microstructure and residual stress of nanoscale depth profiles for TSV-Cu/TiW/SiO$_2$/Si interfaces after different thermal loading, *Materials* 16 (2023) 449.

[24]   J. Xu, C. Cai, V. Pham, K. Pan, H. Wang, and S. Park, A comprehensive study of electromigration in lead-free solder joint, in: *2020 IEEE 70th Electronic Components and Technology Conference (ECTC)*, IEEE, Orlando, FL, 2020: pp. 284–289.

[25]   D. Bani Hani, R. Al Athamneh, M. Abueed, and S. Hamasha, Reliability modeling of the fatigue life of lead-free solder joints at different testing temperatures and load levels using the Arrhenius model, *Scientific Reports* 13 (2023) 2493.

[26]   H.-C. Cheng, C.-L. Ma, and Y.-L. Liu, Development of ANN-based warpage prediction model for FCCSP via subdomain sampling and Taguchi hyperparameter optimization, *Micromachines* 14 (2023) 1325.

Chapter 2

Warpage evolution of TSV wafer based on a hierarchy multiscale analysis method

2.1 Introduction

Through silicon vias (TSVs) represent the core technology for 2.5D/3D packaging. TSVs offer the advantages of achieving the shortest interconnects, smallest pad size and pitch, superior electrical performance, lower power consumption, wider data bandwidth, higher interconnect density, smaller form factor, and lighter weight. The heterogeneous integration of multiple chips using TSV-based interposer boards can enable scaling up similar chips or the high-density integration of chips with various functionalities [1]. Multiple homogeneous or heterogeneous chips are flip-chip mounted onto a TSV interposer, with signal transmission between the chips achieved through re-routing layers and micro bumps [2]. During the entire packaging process of TSV interposers, which is completed on TSV wafers, stress accumulation occurs after processes such as thermal curing and thinning, manifesting macroscopically as complex wafer warping, as illustrated in Figure 2.1. In the TSV wafer fabrication process, warping is particularly pronounced due to the curing of polyimide (PI), bonding, and wafer backside thinning. Therefore, effective control and prediction of warping during the TSV wafer processing are crucial for improving product yield. Wafer warpage not only significantly impacts subsequent packaging processes such as electroplating and dicing, as well as automated operations, but also gives rise to various reliability issues such as microcracks, interfacial delamination, and fragmentations. Therefore, wafer warpage has become one of the critical issues affecting the reliability and yield of TSV wafers.

This chapter first provides a detailed introduction to the process flow and process parameters of TSV wafers. Subsequently, it describes the packaging structure, dimensional parameters of TSV wafers, and the internal structure of TSV interposers. Due to the multiscale structural characteristics of TSV wafers, establishing a comprehensive numerical model presents significant challenges. Therefore, based on the structural features of TSV wafer packaging, this chapter proposes for the first time a hierarchical multiscale analysis method to construct a high-precision finite-element model of TSV wafers and elucidates its analytical process. This offers a novel solution to address the multiscale structural issues in TSV wafers.

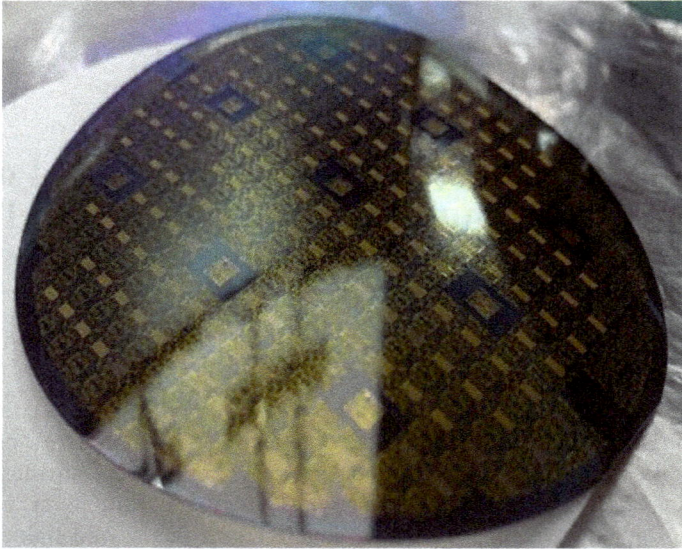

Figure 2.1   Warpage issue of wafer-level packaging

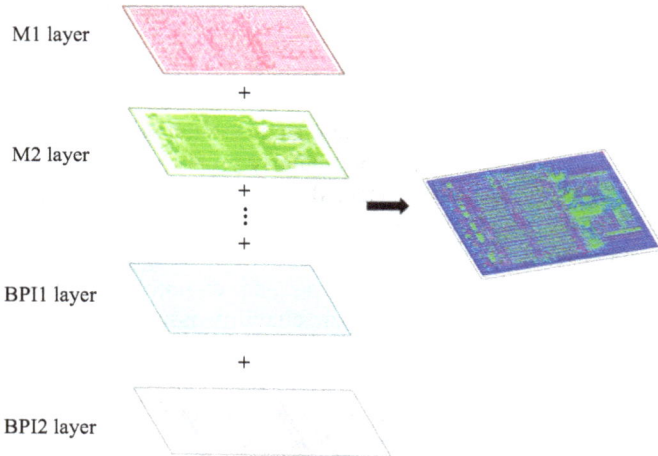

Figure 2.2   The pattern of TSV interposer

2.2   Fabrication process of TSV wafer

Figure 2.2 depicts the layout of a single interposer board, which consists of multiple functional layers including multilayer routing layers, passivation layers, and the interposer layer. In actual production, various processes such as thinning, electro-plating, and curing are required to complete the manufacturing of TSV interposers.

Figure 2.3 illustrates the main process flow of TSV interposer fabrication. Prior to formal processing, incoming wafers are cleaned with plasma water and dried in an oven. After pre-treatment, a 20 μm thick photoresist is spin-coated onto the front side of the wafer, and TSV via patterns are defined through exposure.

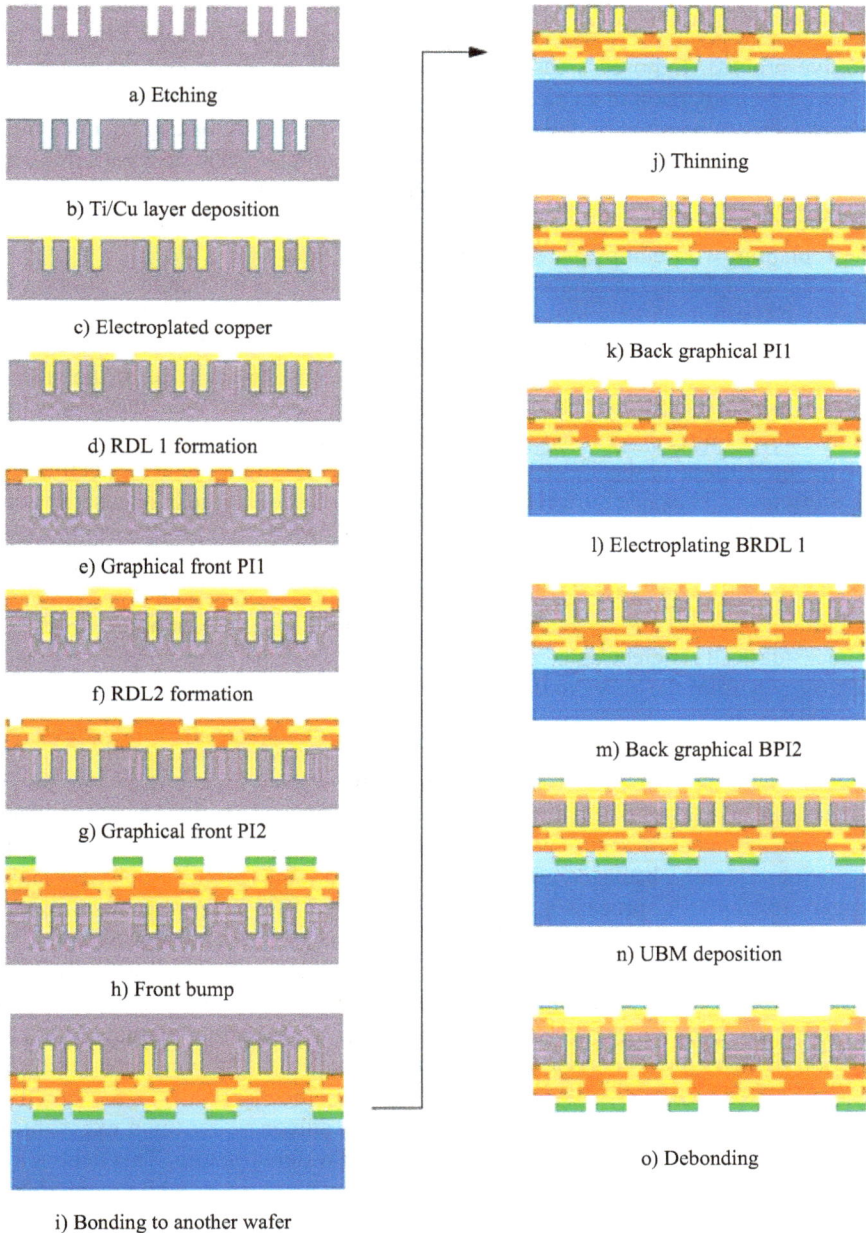

a) Etching

b) Ti/Cu layer deposition

c) Electroplated copper

d) RDL 1 formation

e) Graphical front PI1

f) RDL2 formation

g) Graphical front PI2

h) Front bump

i) Bonding to another wafer

j) Thinning

k) Back graphical PI1

l) Electroplating BRDL 1

m) Back graphical BPI2

n) UBM deposition

o) Debonding

Figure 2.3   Mainly manufacturing process of TSV interposer

Subsequently, TSV blind holes are etched using the Bosch process, with a depth of 200 μm and a diameter of 20 μm. Plasma-enhanced chemical vapor deposition (PECVD) is employed to deposit a 3 μm thick SiO_2 layer on the wafer surface and hole walls, providing insulation. A Ti/Cu metal layer is deposited as a barrier layer to prevent copper–silicon diffusion, with Ti thickness of 500 nm and Cu thickness of 1800 nm. Copper is then electroplated to fill the TSVs at room temperature. Next, the first layer of the redistribution layer (RDL) is electroplated using a semi-additive process. Exposed areas are defined through exposure and development, followed by immersion in an electrolyte bath to electroplate the first layer of RDL. Subsequently, a passivation layer is spin-coated, followed by baking in an oven at 210 °C for curing. The passivation layer serves to relieve stress concentration and provide insulation, similar to the semi-additive process for the front-side second layer of RDL. The final process on the front side involves bump patterning and electroplating, with bumps plated to a height of 32 μm.

After completing the front-side processes, the wafer is bonded to a glass wafer using a 100 μm thick bonding adhesive at a bonding temperature of 220 °C. This adhesive provides mechanical support during subsequent thinning processes. Chemical mechanical polishing is then applied to thin the backside of the silicon wafer, ensuring a uniform thickness and smooth surface. Once the top wafer reaches a remaining thickness of approximately 30 μm, dry etching is performed to expose the bottoms of the TSVs by about 10 μm. This step is crucial for facilitating the subsequent backside processing, allowing for the creation of electrical connections and packaging layers. Next, a 5 μm thick passivation layer is spin-coated onto the backside of the wafer and cured at 210 °C. The passivation layer serves as a protective barrier, ensuring the integrity of the silicon substrate and providing insulation between various conductive layers. Following passivation, the first layer of backside RDL is electroplated onto the wafer, which helps to redistribute electrical signals from the TSVs to the outer edges of the wafer, enabling connection to external circuits.

After the RDL is electroplated, a second passivation layer is applied on the backside using the same process parameters as those used for the front-side passivation. This additional passivation ensures that the backside is fully protected from environmental contaminants and mechanical stresses during further processing steps. Subsequently, backside under-bump metallization (UBM) is patterned, and a copper electroplating process is used to thicken the UBM layer to a 5 μm copper thickness. This layer forms the foundation for subsequent solder bumping or interconnection with external devices. The UBM layer is further enhanced with nickel (Ni) and gold (Au) plating, which serve as protective and conductive layers for the copper, ensuring long-term reliability and preventing oxidation. Laser ablation is then employed to achieve the debonding of the silicon wafer from the glass wafer. A layer of laser-responsive material embedded within the bonding adhesive allows for the precise detachment of the silicon wafer upon exposure to laser light, without causing damage to the delicate wafer structure. This debonding process is critical as it allows for the removal of the glass wafer while maintaining the integrity of the TSV interposer. Finally, after the wafer is diced into individual chips, a single TSV interposer board is obtained, ready for integration into advanced semiconductor packages.

The packaging structure of TSV wafers inherently exhibits multiscale characteristics. On one hand, large-scale features such as the overall wafer layout, bonding adhesive layers, and RDLs require global modeling at a macro-scale. On the other hand, small-scale features, such as the intricate TSVs, micro-level stress distributions, material interfaces, and fine geometries of passivation and UBM layers, necessitate detailed modeling at the micro- or nano-scale. This combination of different scales within the same system makes numerical analysis particularly challenging.

Numerical simulations of such multi-scaled structures typically require the subdivision of the model into an exceedingly large number of elements to capture both global and local behaviors with adequate accuracy. However, this fine meshing increases the computational cost exponentially, leading to longer simulation times and, in some cases, rendering the analysis infeasible. Moreover, conventional single-scale models may struggle to accurately capture the complex interactions between these different scales, leading to potential errors in the prediction of device performance, reliability, and failure modes. To address these challenges, there is an urgent need for the development of multi-scale analysis approaches that can handle both large-scale and small-scale features simultaneously in a computationally efficient manner. Such methods would ideally integrate coarse-scale models for global structural analysis with fine-scale models for localized effects, such as stress concentrations around TSVs, thermal gradients, and material behaviors at the microscopic level. By enabling fast and accurate simulations across multiple scales, these approaches would facilitate more efficient design and optimization of TSV wafer structures, enabling semiconductor manufacturers to streamline their development processes and improve the performance, reliability, and cost-effectiveness of advanced packaging technologies. Furthermore, multi-scale modeling could assist in predicting failure mechanisms more accurately, ultimately contributing to longer-lasting and more robust semiconductor devices.

2.3   Hierarchy multiscale analysis method

As stated earlier, the layout of TSV interposer boards is formed by the stacking of TSV layers, multiple layers of RDL, and PI layers, resulting in a highly intricate physical structure. The positioning of PI openings and RDL traces is determined based on the number of I/O and the location of TSV-Cu. Even with a single layer of RDL, the routing is exceedingly complex. Therefore, it is essential to homogenize these complex structures to establish equivalent mechanical models, thereby reducing the difficulty of modeling and computational scale. In practice, RDL and TSV-Cu are commonly encountered in wafer-level packaging structures, posing significant challenges to the industry. This section proposes a hierarchical multiscale analysis approach to address the multiscale issues inherent in TSV wafer packaging structures.

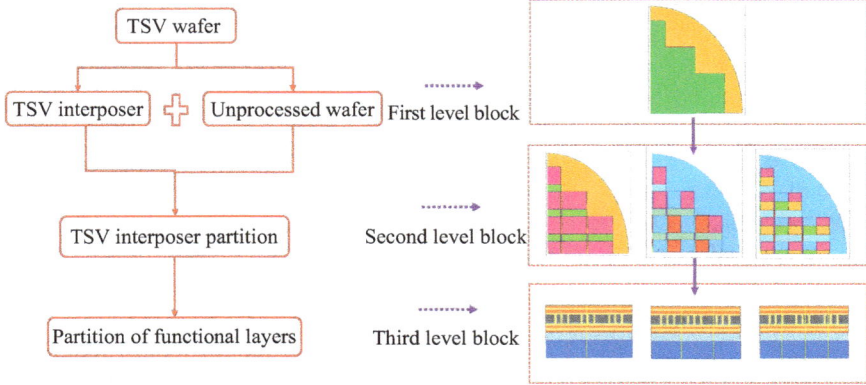

Figure 2.4   Analysis process of hierarchical multiscale method

The core concept of hierarchical multiscale analysis lies in categorizing and partitioning TSV packaging structures according to their structural characteristics, and replacing the actual model with homogenized models derived from these categories and partitions, thereby establishing high-precision finite-element models. The computational modeling process of hierarchical multiscale analysis is illustrated in Figure 2.4. This process divides the TSV wafer structure into three levels based on its characteristics, each level referred to as a block. The first-level block separates the functional areas of the TSV interposer board from the unprocessed wafer. Given that the TSV interposer board is designed with one graphic area and three test areas, it can be subdivided into four regions, constituting the second-level block. Previous study [3] has equivalently represented the PI layer and even disregarded the impact of the RDL layer when investigating wafer warpage evolution. Building upon the second-level block, this chapter introduces the third-level block, which similarly partitions different functional layers. Upon completing all partitions, the effective material parameters for each homogenized region are computed based on the composition of materials in different areas. These parameters are then inputted into the homogenized model to establish a high-precision equivalent model of the TSV wafer. Theoretically, a higher number of subdivisions at the lowest level results in a model that closely approximates the real structure. When the subdivision regions precisely match the actual packaging structure, the model established represents the true packaging structure. The effective material parameters for the homogenized regions in the linear phase are derived using modified composite material mechanics formulas [4–6], as illustrated in the following equations:

$$\begin{cases} E_z = V_d E_d + V_m E_m \\ E_x = E_y = E_d E_m / (V_m E_d + V_d E_m) \end{cases} \tag{2.1}$$

where E is the elastic modulus, V_m represents the volume fraction of copper in a single region, V_f represents the volume fraction of PI in a single region, m is copper, and d represents silicon or PI

$$\begin{cases} \mu_{xz} = \mu_{yz} = E_x(V_d\mu_d + V_m\mu_m)/E_z \\ \mu_{xy} = V_m\mu_m + V_d\mu_d\left[1 + \mu_d - \left(\dfrac{E_d}{E_d}\right)\mu_{xz}\right] \bigg/ \left[1 - u_d^2 + \mu_d\left(\dfrac{E_d}{E_z}\right)\mu_{xz}\right] \end{cases} \quad (2.2)$$

where v_m is the Poisson's ratio of copper, v_f represents the Poisson's ration of silicon or PI, subscript xy is the plane xy, subscript yz is the plane yz, and subscript xz is the plane xz

$$\begin{cases} G_{xz} = G_{yz} = G_dG_m/(V_mG_d + V_dG_m) \\ G_{xy} = E_x/\left[2\left(1 + \mu_{xy}\right)\right] \end{cases} \quad (2.3)$$

where G_m is the shear modulus of copper and G_f represents the shear modulus of silicon or PI

$$\begin{cases} \alpha_z = [V_d\alpha_dE_d + V_m\alpha_mE_m]/E_z \\ \alpha_x = \alpha_y = \alpha_dV_d(1 + \mu_d) + \alpha_mV_m(1 + \mu_m) - \alpha_z\mu_{xz} \end{cases} \quad (2.4)$$

where α_m is the CTE of copper and α_f represents the CTE of silicon of PI.

2.4   Numerical simulation of TSV wafer warpage

2.4.1   Packaging structure

The subject of this study is an 8-in. TSV wafer. As shown in Figure 2.5, the front side of the wafer contains 23 interposers. Due to their rectangular geometry, these interposer boards do not occupy the entire wafer, leaving unprocessed areas around the TSV wafer perimeter. Each individual interposer board is divided into four regions, with the largest area located in the upper right quadrant designated as the

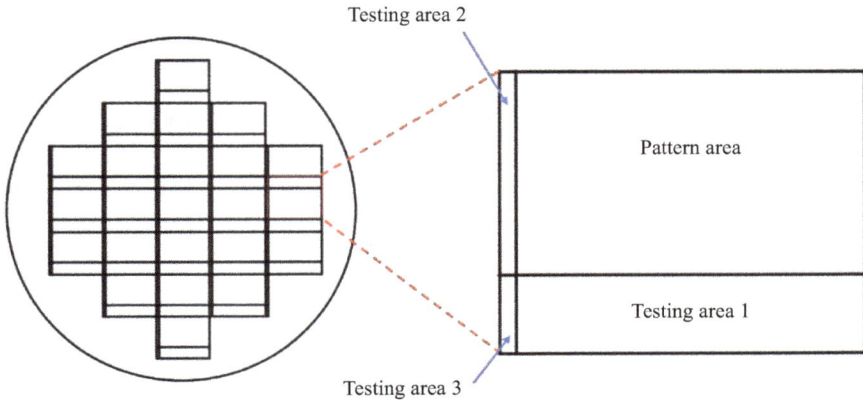

Figure 2.5   Dimension and layout of TSV interposer on an 8 in. wafer

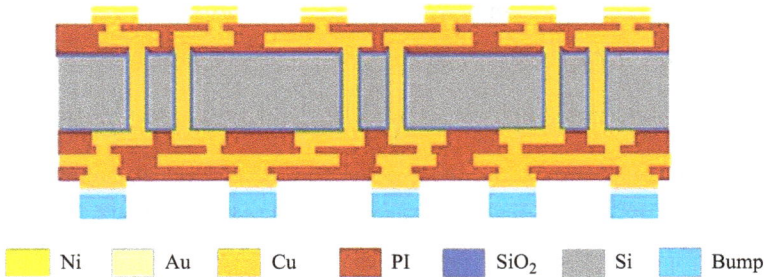

Figure 2.6   *The cross-section of TSV interposer*

Table 2.1   *Dimensions of packaging structure*

Component	Geometry size
Wafer diameter	8 in.
Wafer thickness	730 μm
Interposer dimensions (length × width)	32.66 mm × 25.16 mm
Thickness of PI1	5 μm
Thickness of M1	5 μm
Thickness of M2	5 μm
Thickness of PI2	15 μm
Thickness of TSV	200 μm
Thickness of SiO_2	2.5 μm
Thickness of BPI1	15 μm
Thickness of BPI2	15 μm
Thickness of BM1	5 μm

graphic area, while the remaining areas are referred to as test area 1, test area 2, and test area 3, respectively.

The cross-sectional view of an individual TSV interposer board, obtained after die bonding and dicing, is depicted in Figure 2.6. For clarity in description, the front-facing layers are designated as RDL1 and RDL2 for the first and second layers of RDL, respectively, while the back-facing first layer of RDL is denoted as BRDL1. Similarly, the front-facing layers of PI are named PI1 and PI2, and the back-facing layers as BPI1 and BPI2. Thus, from top to bottom, the interposer board structure comprises metal bumps (Bump), UBM, PI2 layer, RDL2 layer, PI1 layer, RDL1 layer, SiO_2 layer, TSV layer, BPI1 layer, BRDL1 layer, BPI2 layer, and UBM.

2.4.2   Package structural parameters

The size parameters of the TSV wafer and interposer boards are detailed in Table 2.1, provided by our collaborators. From these structural parameters, it is evident that the TSV wafer packaging structure spans from millimeter-scale

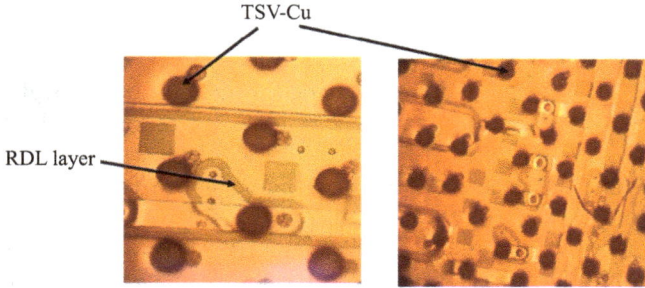

Figure 2.7   The pattern of TSV interposer before polishing

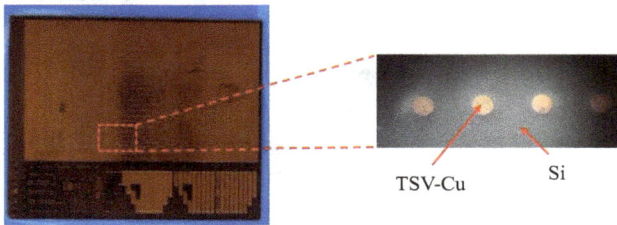

Figure 2.8   The sample of TSV interposer

dimensions such as the wafer diameter to micrometer-scale dimensions like TSV-Cu diameter and RDL linewidth. The entire structure exhibits multiscale characteristics, ranging in feature size from micrometers to millimeters. Additionally, the interposer boards contain numerous TSV-Cu structures that are unevenly distributed and feature complex routing layers.

Figure 2.7 displays the morphology of the interposer board under an optical microscope before polishing, revealing oxidized TSV-Cu where the black areas indicate copper oxidation. Figure 2.8 presents a physical image of the TSV interposer board after dicing. The interposer board samples are encapsulated with two layers of blue oxide film to delay metal oxidation, as shown on the left side of Figure 2.8. To observe the internal structure of the interposer board, it was cut for sample preparation. After embedding the sample in epoxy resin, rough grinding was performed using 800-grit sandpaper. Subsequently, the rough-ground interposer board sample was polished using a dual-disc dual-control metallographic polishing machine with 1 μm diamond suspension to remove the PI and RDL layers from the surface, thereby exposing the TSV-Cu. The right side of Figure 2.8 shows a partial area of the polished interposer board sample under an optical microscope, where the gray areas represent the silicon substrate and the light-yellow areas denote electroplated TSV-Cu, crucial for chip-to-substrate signal transmission and thermal dissipation channels.

2.4.3   Acquisition of material parameters and establishment of FE model

The TSV wafer is segmented into various homogenized regions through hierarchical multiscale analysis. To derive equivalent material parameters, the layout of these regions is transformed into a binary map. This two-dimensional matrix exclusively comprises 0 and 1, corresponding to the two-phase materials in the plate image. By tallying the occurrences of 0 and 1, the composition of each functional layer's materials is determined. Figure 2.9 illustrates the conversion process from the RDL1 layer layout image to a binary map. Similar methods are employed to ascertain the proportion of different material components in other homogenized regions.

Table 2.2 presents the copper content for each functional layer of the interposer when divided into four equivalent regions. Once the copper content rates for each layer are determined, the segmented homogenized regions are treated as transversely

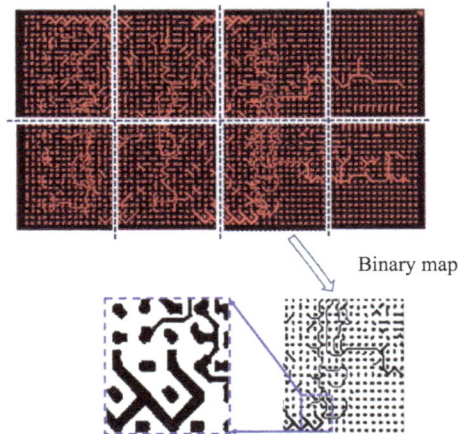

Binary map

Figure 2.9   Extraction of copper proportion in different function layers of TSV interposer

Table 2.2   TSV interposer functional layer and copper content in each area

	Pattern area (%)	Testing area 1 (%)	Testing area 2 (%)	Testing area 3 (%)
TSV layer	0.78	0.74	0.73	0.35
M1 layer	77.23	77.23	11.20	11.20
PI1 layer	0.48	1.20	0.60	2.10
M2 layer	24.43	16.00	11.20	6.00
PI2 layer	3.76	5.00	8.85	5.70
BPI layer	0.14	1.00	2.00	2.00
BM1 layer	24.31	13.00	11.20	11.00
BPI2 layer	9.58	10.00	5.30	5.30

Table 2.3   *Equivalent material properties*

	Direction	Elastic modulus (GPa)	Poisson's ratio	Shear modulus (GPa)	CTE (ppm/°C)
TSV layer	z	130.99	0.30	50.33	2.90
	x,y	130.88	0.30	50.32	2.92
M1 layer	z	90.93	0.04	3.98	17.23
	x,y	10.24	0.36	3.77	32.82
PI1 layer	z	3.04	0.23	0.98	47.25
	x,y	2.51	0.31	0.96	57.92
M2 layer	z	30.47	0.03	1.28	19.29
	x,y	3.29	0.38	1.19	57.22
PI2 layer	z	6.80	0.11	1.01	30.09
	x,y	2.60	0.37	0.95	64.14
BPI1 layer	z	2.66	0.26	0.98	51.72
	x,y	2.50	0.29	0.97	55.42
BM1 layer	z	30.33	0.03	1.28	19.31
	x,y	3.28	0.38	1.19	22.31
BPI2 layer	z	13.47	0.06	1.08	22.31
	x,y	2.76	0.38	1.00	63.33

isotropic materials. Using (2.1) to (2.4), the mechanical parameters of the two-phase composite material in its elastic phase are calculated. Table 2.3 displays the equivalent material parameters for the elastic phase within the homogenized regions.

In addition to obtaining equivalent material parameters for the elastic phase of the homogenized regions, parameters for the plastic phase of these regions are extracted based on the developed RVE method coupled with the generalized Hill plasticity model, completing the equivalency and validation of the critical model. The stress–strain behavior of TSV-Cu parameters from the literature [7] under non-annealed conditions. The obtained elastoplastic parameters are collectively assigned as inputs to the established homogenized model, replacing the actual model within the TSV wafer, thereby establishing a high-precision finite-element model of the TSV wafer.

To reduce computational scale and enhance simulation efficiency, a quarter model is established based on the symmetry of the packaging structure. The TSV wafer is analyzed using a hierarchical multiscale analysis approach, wherein four models are created corresponding to different numbers of segmented regions in the interposer layout: 1, 2, 4, and 8.

Figure 2.10 depicts a quarter finite-element model of the interposer, divided into four equivalent regions using ANSYS 18.0. The element type employed is Solid185, with a total of 844,672 elements. Due to the large number of elements and nodes in the model, computational time is minimized by utilizing computing resources from the Beijing Super Cloud Computing Center (AMD 7452 @ 2.35 GHz, 128 cores, 256 GB memory) for modeling and simulations (Table 2.4).

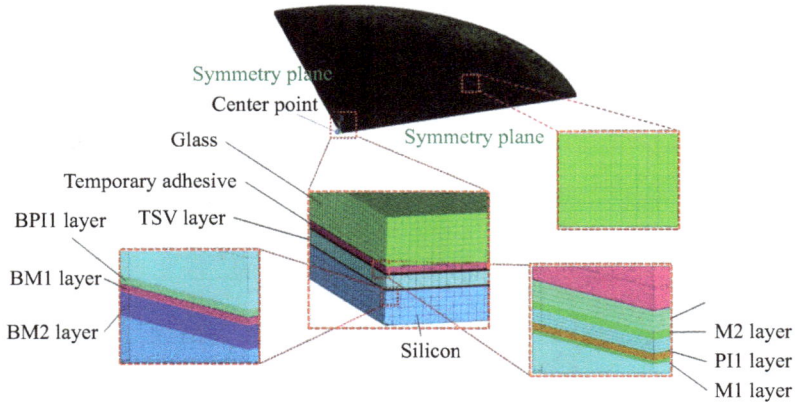

Figure 2.10   Quarter finite-element model

Table 2.4   Material properties utilized in simulation [7–10]

Material	Elastic modulus (GPa)	Poisson's ratio	CTE(ppm/°C)	T_g (°C)
Glass	73.6	0.23	3.5	–
Copper	117	0.35	17.0	–
Silicon	131	0.30	2.8	–
SiO$_2$	76.7	0.20	0.6	–
Temporary adhesive	1.2(E1)/0.4(E2)	0.38	45(CTE1)/90 (CTE2)	120
PI	2.5	0.28	54.0	–

Symmetry boundary conditions are applied to the symmetry plane of the quarter FE model, with full constraints imposed on the central node position along the bottom edge of the symmetry plane to restrict rigid body displacements across the entire model. TSV wafers undergo a series of processes during the flow wafering process. To address the impact of these processes on wafer warpage, the flow wafering process is segmented into 27 distinct steps. The process parameters for each step are applied to the model as loads to simulate real-world conditions comprehensively. The loading temperature curve is illustrated in Figure 2.11, with specific temperatures determined based on process conditions.

The simulation workflow utilizing the birth and death element technique based on the hierarchical multi-scale model is illustrated in Figure 2.12. Initially, the overall model is established. Next, process parameters are applied as loads to the model. Subsequently, material addition and removal are performed using the birth and death technique to simulate the process. Upon completion of the simulation, results are compared with experimental data. If discrepancies are significant, the model is revised and recalculated accordingly. Conversely, alignment of warpage

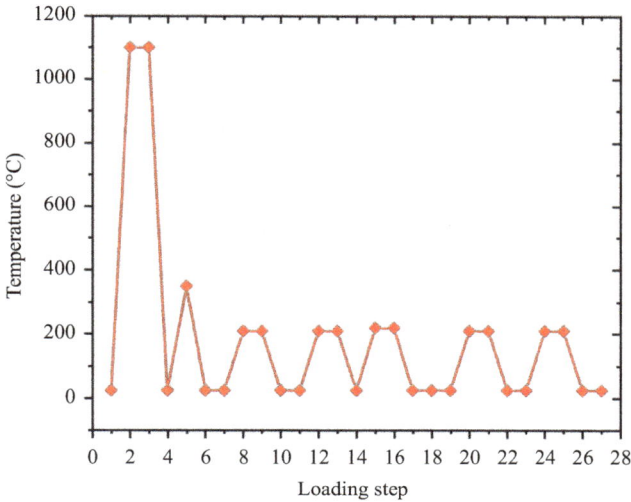

Figure 2.11   Process temperature corresponding to each load step

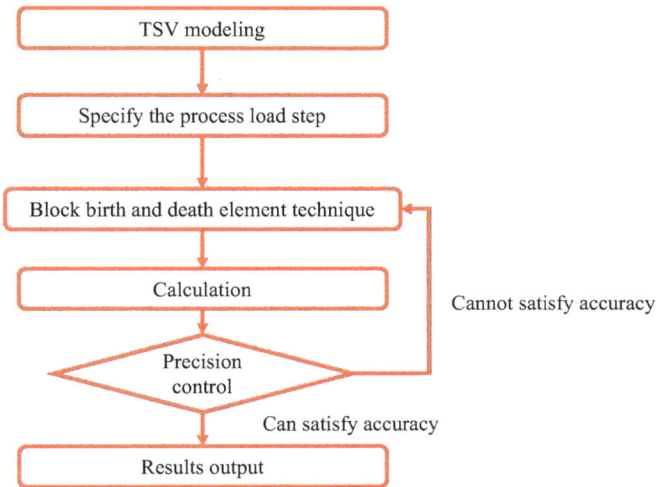

Figure 2.12   Simulation diagram of element birth and death technology based on hierarchical multiscale method

values and directions with experimental results indicates the model's ability to guide subsequent processes effectively.

Table 2.5 presents the specific meanings of each load step corresponding to the processes involved. The simulation process employs a layer-by-layer "kill" and "activate" approach to model the warpage evolution of TSV wafers during process fabrication. Initially, a complete model including glass wafers, bonding adhesives, and

Table 2.5   Process corresponding to load step

Load step	Processing	Load step	Processing
1	Kill glass, PI1, PI2, and M2 layer element	14	Cooling to room temperature (25 °C)
2	Heating to 1100 °C	15	Heating to 220 °C (bonding temperature)
3	Activate SiO_2 layer element	16	Activating adhesive and glass layer element
4	Cooling to room temperature (25 °C)	17	Cooling to room temperature (25 °C)
5	Heating to 350 °C (annealing temperature)	18	Killing silicon element on the backside of wafer
6	Cooling to room temperature (25 °C)	19	Killing silicon elements on the backside of the wafer to the TSV layer
7	Activate M1 layer element	20	Heating to 210 °C (PI curing temperature)
8	Heating to 210 °C (PI curing temperature)	21	Activating BPI1 layer element and replacing materials
9	Active PI1 layer	22	Cooling to room temperature (25 °C)
10	Cooling to room temperature (25 °C)	23	Activating BM2 layer element and replacing materials
11	Activate M2 layer element	24	Heating to 210 °C (PI curing temperature)
12	Heating to 210 °C (PI curing temperature)	25	Activating BPI2 layer element and replacing materials
13	Active PI2 layer	26	Cooling to room temperature (25 °C)
		27	Killing adhesive and glass wafer elements

functional layers is established. Structures not yet fabricated are "killed" as the initial state for simulation. Subsequently, the TSV layer is "activated" to simulate the TSV electroplating process, followed by a global heating to 350 °C and subsequent cooling to room temperature to mimic TSV annealing. The wafer is then heated from room temperature to 1100 °C, "activating" the SiO_2 layer, and subsequently cooled to simulate PECVD processing. After completion, the M1 layer is "activated" to simulate RDL1 electroplating. The temperature is raised to the process temperature of PI at 210 °C, "activating" PI1 layer units, followed by cooling to room temperature to complete PI1. The same method is used to simulate the RDL2 electroplating process. Processes such as Ti/Cu layer deposition, bump formation, dicing, which minimally affect warpage, are omitted. Thus, the simulation of the TSV wafer's front-side processes is completed. For back-side processes, the model is first heated to 220 °C and the glass wafer layer is "activated," followed by cooling to room temperature to simulate bonding processes. Back-side thinning involves "killing" silicon units until TSVs are exposed. BPI1 curing is simulated by heating the model to a curing temperature of 210 °C, then "activating" BPI1 layer units using the MPCHG command to replace silicon material with PI material, followed by cooling to room temperature. BRDL1 layer units are activated and material replacement completed to simulate back-side electroplating processes. Upon completion, the model is heated again to PI adhesive curing temperature, followed by "activating" BPI1 layer units to simulate BPI1 fabrication and cooling to room temperature to complete back-side wafer processes. Processes for back-side PI and RDL layers are similar to those for the front side, with the final simulation involving "killing" bonding adhesives and glass carriers to simulate debonding processes. It is noteworthy that unlike the front-side simulation, material replacement using the MPCHG command is required during the simulation of BPI1, BPI2 curing, and BRDL electroplating on the back side. Other processes such as Ti/Cu layer deposition, TSV patterning, bump formation, UBM patterning, and dicing, which have minimal effect on warpage, are not listed.

2.4.4   Results and discussion

Based on the hierarchical multi-scale approach, an equivalent model of TSV wafer packaging structure is established to simulate the wafer warpage evolution during the process. The entire process is divided into 27 load steps. Figure 2.13 illustrates the trend of maximum wafer warpage values throughout the process and the wafer warpage contour map after debonding.

It can be observed that the direction of wafer warpage alternates with the process stages. The maximum warpage occurs during the bonding process, reaching 300 μm with a "smile" shape. As the process progresses, the warpage direction gradually changes, and after debonding, the wafer exhibits a "frown" direction with a maximum warpage of −282.75 μm. From the initial steps to bonding, all processes occur on the wafer's front side, sequentially "activating" units previously "killed," resulting

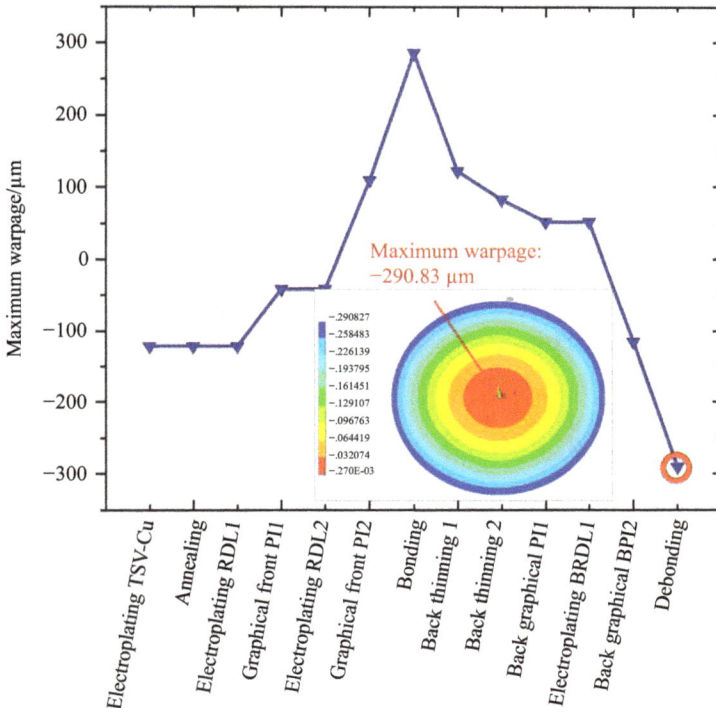

Figure 2.13   TSV wafer warpage evolution during the process

in consistent warpage direction. However, post-debonding processes are conducted on the wafer's back side, where reverse stresses from back-side processes continuously reduce the maximum wafer warpage until it exhibits reverse warpage.

Figure 2.14 depicts the warpage contour of the wafer upon completion of the primary process stages. In Figure 2.14a, the simulated warpage after plating the

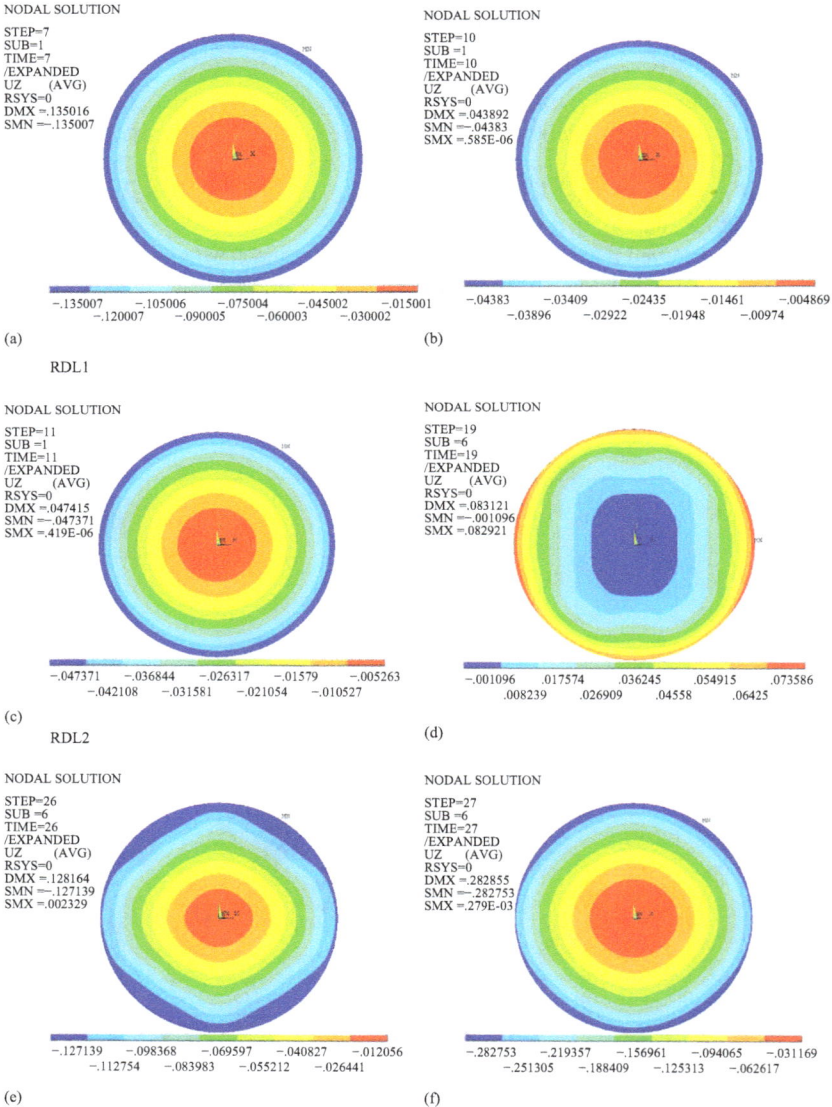

Figure 2.14   Wafer warpage morphology after different processes: (a) Warpage contour after electroplating, (b) Warpage contour after PI1 process, (c) Warpage contour after electroplating, (d) Warpage contour after back grinding, (e) Warpage contour after BPI2 process, and (f) Wafer warpage morphology after debonding process

RDL1 layer shows a maximum warp of −135.00 μm. During the PI1 fabrication, the wafer experiences an overall temperature rise to 210 °C (curing temperature of PI), subsequently cooled to room temperature, resulting in a warpage value of −43.83 μm as illustrated in Figure 2.14b. Activation of RDL2 layer elements for room temperature plating shows a marginal effect on wafer warpage, evidenced by a final warpage of −47.37 μm depicted in Figure 2.14c. Upon completing RDL2 plating, the temperature is raised to 350 °C, followed by cooling to room temperature to finalize PI2, yielding a wafer warpage of 116.51 μm. The curing process of PI induces shrinkage, consequently increasing the warpage. Subsequent bonding of the wafer to a glass carrier board on the front side, followed by back-side processing, involves raising the temperature to 220 °C for bonding, activating bonding adhesive and glass carrier layer elements, and cooling to room temperature to complete bonding, resulting in a warpage of 282.96 μm.

After completion of bonding, silicon elements undergo simulated wafer thinning. During this process, the wafer gradually decreases in thickness, culminating in a maximum warpage value of 82.92 μm, as depicted in Figure 2.14d. Subsequently, the backside processing continues until the TSV-Cu is exposed. To simulate the BPI1 process, the entire model is heated to the curing temperature of the polyimide (PI), coinciding with material replacement, resulting in a warpage value of 50.28 μm. Maintaining the temperature constant to activate the M1 layer elements and facilitate material replacement, minimal variation in warpage is observed. BPI2 is then produced under analogous process conditions to other PI layers, resulting in a maximum warpage value of −127.14 μm upon completion, illustrated in Figure 2.14e.

After completing the wafer flow process, the FSM 500TC was employed to assess the wafer's warpage morphology. As depicted in Figure 2.15, the measured

BOW HEIGHT ALONG THE SCAN PATH

Wafer 1, Scan 1, laser.780, casselle: Noname_2019_08_14_11_45_50

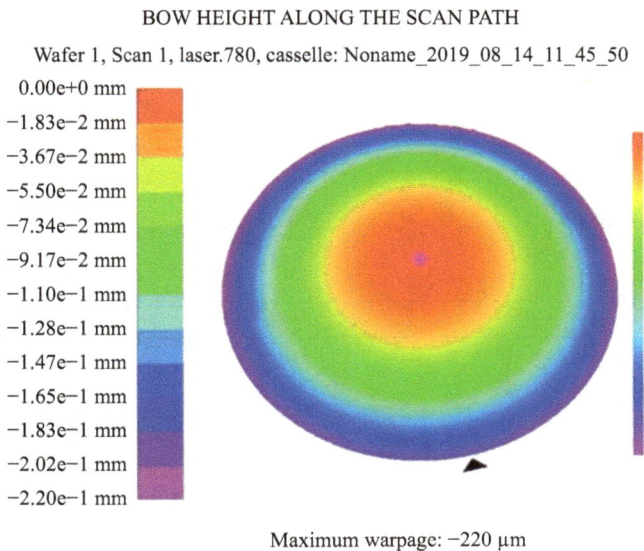

| 0.00e+0 mm |
| −1.83e−2 mm |
| −3.67e−2 mm |
| −5.50e−2 mm |
| −7.34e−2 mm |
| −9.17e−2 mm |
| −1.10e−1 mm |
| −1.28e−1 mm |
| −1.47e−1 mm |
| −1.65e−1 mm |
| −1.83e−1 mm |
| −2.02e−1 mm |
| −2.20e−1 mm |

Maximum warpage: −220 μm

Figure 2.15　Wafer warpage measurement after debonding process

post-debonding wafer exhibits a characteristic "crying face" warpage profile, with a maximum value of −220 µm. This observed warpage tendency aligns closely with the numerical simulation results illustrated in Figure 2.14f, which also depicts a "crying face" shape, showcasing a maximum warpage value of −282.75 µm. The discrepancy between these values is merely 62.75 µm.

Figure 2.16 illustrates the evolution of wafer warpage predicted through the hierarchical multiscale method. The maximum warpage of the wafer, 282.96 µm, is observed during the bonding process, contrasting with the previous assumption that maximum warpage occurs at the process's conclusion. Therefore, meticulous attention to wafer warpage post-bonding is crucial, as larger warpage values can significantly impact subsequent process quality, ultimately reducing overall yield. Furthermore, the influence of subdivision number on final wafer warpage can be discerned; increasing subdivisions progressively aligns simulated wafer warpage values closer to −220 µm, reflecting improved representation of RDL and TSV-Cu distribution heterogeneities in the equivalent model, converging toward real-world conditions. Figure 2.17 illustrates the effect of subdivision number on wafer warpage in the interposer pattern area, reducing calculation error from 28.52% to 10.45% when employing hierarchical multiscale modeling with experimental measurements as the benchmark. Notably, due to partition limitations, the

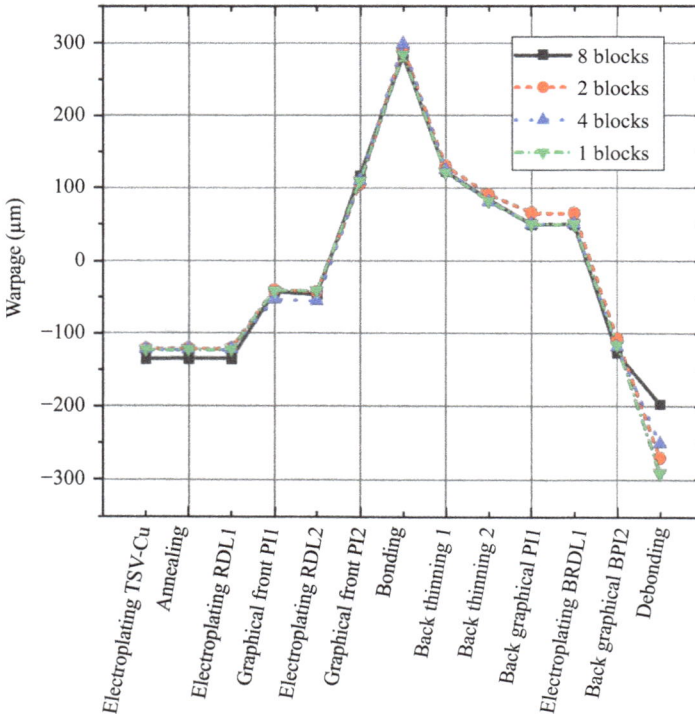

Figure 2.16   Effect of the number of partitions on wafer warpage evolution

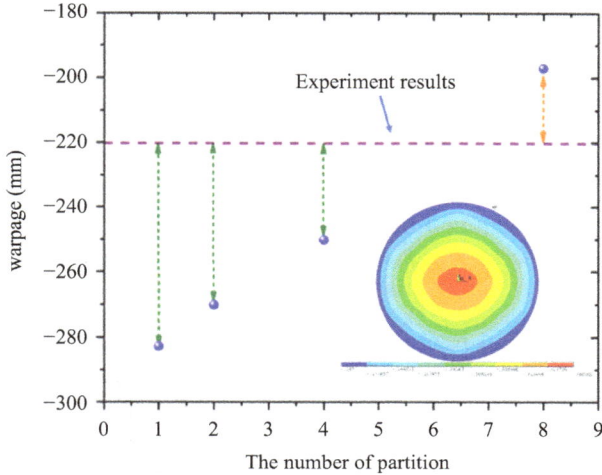

Figure 2.17   Impact of the number of partitions on prediction accuracy

established equivalent model exhibits slight deviations compared to experimental results.

The number of elements in the equivalent model, subdividing the interposer pattern area into 2, 4, and 8 partitions, are 842,959, 858,090, and 860,712, respectively. This minor variation indicates that as the number of partitions increases, the equivalent model more closely approximates the real model. Consequently, the discrepancy between simulation and experimental results diminishes progressively. The hierarchical multiscale method proposed in this study is adaptable and enhances simulation accuracy for addressing multiscale challenges in package structures.

2.5   Chapter summary

The hierarchical multiscale method proposed in this study is utilized to develop a highly accurate and robust prediction model for the warpage behavior of TSV wafers during the entire encapsulation process. This equivalent model is capable of predicting the evolution of wafer warpage with remarkable precision, encompassing every stage of the TSV wafer packaging process, from initial bonding to final debonding. In addition to this, the study systematically explores the impact of partitioning the interposer pattern area into multiple zones, evaluating how different partition configurations affect the accuracy of the warpage predictions. This leads to the formulation of an optimization strategy for enhancing model performance. The key findings from the study can be summarized as follows:

The study reveals that wafer warpage exhibits significant variations in both magnitude and direction at different stages of the TSV packaging process. Notably, the largest observed warpage, measuring -282.96 μm, occurs during the bonding stage of encapsulation, rather than at the final stage of the process as previously expected. This highlights the complexity of wafer warpage dynamics and emphasizes the importance of continuous monitoring throughout each phase of the process. Ensuring wafer integrity requires careful attention to warpage values at every step, rather than solely focusing on the end point.

To further enhance the prediction accuracy, the interposer's pattern area was divided into varying numbers of zones—1, 2, 4, and 8—allowing the creation of corresponding equivalent models for each partition configuration. The results from these partitioned models showed consistent trends in the evolution of wafer warpage, with the step at which the maximum warpage occurs remaining unchanged across different partitions. However, as the number of partitions increased, the prediction accuracy improved. When the interposer was divided into eight partitions, the error in the predicted warpage was reduced to just 10.45%. This demonstrates that using a higher number of partitions significantly enhances the accuracy of the warpage prediction, making it a highly effective strategy for improving model precision.

The findings underscore the effectiveness of the hierarchical multiscale method-based equivalent model in capturing the full spectrum of wafer warpage behavior throughout the TSV packaging process. This model serves as a valuable tool for guiding the management of wafer warpage in actual TSV wafer production lines. Minimizing wafer warpage is crucial not only for improving the yield of the final product but also for reducing production costs, as excessive warpage can lead to defects, reduced reliability, and increased manufacturing time. By integrating the optimized partitioning strategy and carefully monitoring the warpage throughout the packaging stages, manufacturers can enhance the overall quality and cost-effectiveness of their TSV wafer processes.

In conclusion, this study provides a comprehensive approach to simulating and managing wafer warpage in TSV wafer packaging. By leveraging advanced prediction models and optimization strategies, the approach offers substantial potential for improving both the design and manufacturing of TSV-based electronic devices, ensuring their performance, reliability, and economic viability in the highly competitive semiconductor industry.

References

[1]   X. Zhang, J.K. Lin, S. Wickramanayaka, *et al.*, Heterogeneous 2.5D integration on through silicon interposer, *Applied Physics Reviews* 2 (2015) 021308.

[2]   G. Katti, S.W. Ho, Li Hong Yu, *et al.*, Fabrication and assembly of Cu-RDL-based 2.5-D low-cost through silicon interposer (LC–TSI), *IEEE Design & Test* 32 (2015) 23–31.

[3]   J.H. Lau, M. Li, L. Yang, *et al.*, Warpage measurements and characterizations of fan-out wafer-level packaging with large chips and multiple redistributed layers, *IEEE Transactions on Components, Packaging and Manufacturing Technology* 8 (2018) 1729–1737.

[4]   P.K. Mallick, *Fiber-Reinforced Composites: Materials, Manufacturing, and Design,* Third Edition, CRC Press, Boca Raton, FL, 2007.

[5]   G. Hu, Y. Goh Kim, and L. Judy, Micromechanical analysis of copper trace in printed circuit boards, *Microelectronics Reliability* 51 (2011) 416–424.

[6]   L. McCaslin and S.K. Sitaraman, Methodology for modeling substrate warpage using copper trace pattern implementation, in: *2008 58th Electronic Components and Technology Conference,* IEEE, Lake Buena Vista, FL, 2008: pp. 1582–1586.

[7]   Y. Li, P. Chen, F. Qin, *et al.,* Constitutive modelling of annealing behavior in through silicon vias-copper, *Materials Characterization* 179 (2021) 111359.

[8]   K. Wakamoto, Y. Mochizuki, T. Otsuka, K. Nakahara, and T. Namazu, Temperature dependence on tensile mechanical properties of sintered silver film, *Materials* 13 (2020) 4061.

[9]   G. Yang, Z. Kuang, H. Lai, *et al.,* A quantitative model to understand the effect of gravity on the warpage of fan-out panel-level packaging, *IEEE Transactions on Components, Packaging and Manufacturing Technology* 11 (2021) 2022–2030.

[10] Z. Chen, Z. Zhang, F. Dong, S. Liu, and L. Liu, A hybrid finite element modeling: artificial neural network approach for predicting solder joint fatigue life in wafer-level chip scale packages, *Journal of Electronic Packaging* 143 (2021) 011001.

Chapter 3

Investigation on fatigue life of solder joints for 2.5D packaging structures

3.1 Introduction

The reliability of solder joints is crucial for the normal operation of 2.5D packaging structures [1]. The interconnection between chips and substrates relies entirely on the quality and stability of these solder joints. Well-performing solder joints ensure that the packaging structure remains free from connectivity failures during prolonged operation, thereby effectively preventing potential electrical or mechanical faults. Currently, predicting solder joint lifespan poses a significant challenge, especially given the micrometer-level size of solder joints in contrast to the centimeter-level size of substrates in 2.5D packaging. This multiscale nature presents substantial challenges in numerical modeling. To address this, academia is actively exploring and developing homogenization methods to enhance computational efficiency and effectively resolve these complexities. These efforts aim to improve the predictive accuracy of solder joint lifespans in 2.5D packaging structures, ensuring stable operation in various demanding applications.

This chapter first presents the theoretical foundations of the representative volume element (RVE) method and the generalized Hill plasticity model. Building on this, a coupling algorithm based on the RVE method and the generalized Hill plasticity model is developed, and the computational process of the coupling algorithm is outlined. The core idea of this algorithm is to replace the RVE model with a homogeneous model that possesses anisotropic plasticity, ensuring that both models have the same geometric dimensions while the material parameters of the homogeneous model are derived from the RVE model. Subsequently, the proposed method is employed to establish detailed and equivalent models of 2.5D packaging structures. A comparison of stress, strain, displacement, and fatigue life results at critical locations under thermal cycling loads between the two models is performed to validate the effectiveness of the coupling algorithm. The comparative results demonstrate that the coupling algorithm based on the RVE method and the generalized Hill plasticity model accurately describes the mechanical behavior of critical models and can be used for predicting the fatigue life of solder joints in 2.5D packaging structures under thermal cycling conditions.

3.2   Theoretical foundations and analytical process of coupling algorithm

3.2.1   RVE method

The RVE method is suitable for structures with periodic characteristics. This approach can preserve the features of the microscopic model while describing the mechanical behavior of the macroscopic model, making it an effective method to connect micro- and macro-scale models. In the micromechanics of composite materials, many models exhibit periodic structural features, hence this method has been widely applied in establishing numerical models of composite materials [2,3].

The solder joints in 2.5D packaging structures exhibit periodic distribution characteristics, thus allowing for modeling and analysis using the RVE method. Figure 3.1 illustrates the selection of the RVE model and the establishment process of the equivalent model for micro-bump layers. The micro-bump layer is divided into multiple sub-regions, each defined as an RVE model. Material parameters for each RVE model are computed individually, and during modeling, these segmented RVE models are assembled to form the complete model. Each region of the overall model is assigned distinct material parameters. This approach avoids the grid partitioning difficulties associated with establishing microscopic models while closely approximating the real model. Segmenting the overall model and establishing RVE models within it form the basis of applying the RVE method. Boundary conditions applied to the RVE model typically include displacement, force, and periodic boundary conditions. Unlike the other two types, applying periodic boundary conditions ensures coordinated deformation of individual RVEs, preventing

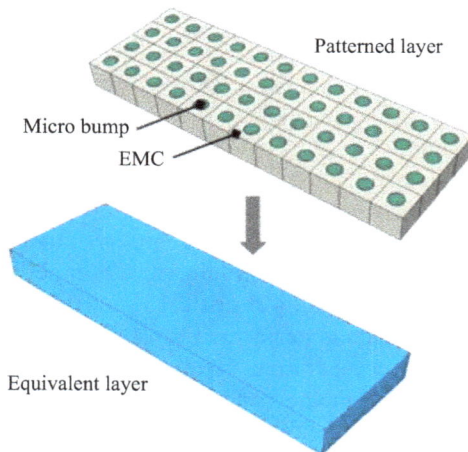

Figure 3.1   Homogenization of micro-bump layer

separation at adjacent RVE interfaces and ensuring stress continuity across these interfaces.

Regarding the description of periodic boundary conditions, Suquet [4] earlier proposed an expression for the periodic displacement field

$$u_i = \bar{\varepsilon}_{ik} x_k + u_i^* \tag{3.1}$$

where $\bar{\varepsilon}_{ik}$ is the average strain, x_k represent coordinates at any arbitrary position, and u_i^* is the displacement correction term determined by applied loads.

Due to the dependence of the parameter u_i^* on applied loads in (3.1), this form is not convenient for applying periodic boundary conditions to the RVE model. Building upon this, Xia *et al.* [5] expressed the nodal displacements on opposite faces of the RVE model as

$$u_i^{j+} = \bar{\varepsilon}_{ik} x_k^{j+} + u_i^* \tag{3.2}$$

$$u_i^{j-} = \bar{\varepsilon}_{ik} x_k^{j-} + u_i^* \tag{3.3}$$

where j^+ means along the positive direction of a coordinate axis and j^- is along the negative direction of a coordinate axis.

The subtraction of the two equations yields

$$u_i^{j+} - u_i^{j-} = \bar{\varepsilon}_{ik} \left(x_k^{j+} - x_k^{j-} \right) = \bar{\varepsilon}_{ik} \Delta x_k^j \tag{3.4}$$

For the specified RVE model, Δx_k^j representing the length of the RVE model in the j-direction. Once strain $\bar{\varepsilon}_{ik}$ is determined, the right-hand side of (3.4) becomes constant. Therefore, (3.4) can be rewritten as

$$u_i^{j+}(x,y,z) - u_i^{j-}(x,y,z) = c_i^j (i,j = 1,2,3) \tag{3.5}$$

where c_1^1, c_2^2, and c_3^3 represent displacements in the directions x, y, and z, and c_1^2, c_2^3, and c_2^3 are shear deformations in three planes xy, yz, and yz.

Due to the derivation, (3.5) does not include correction terms that need to be determined by the load; hence, it is straightforward to apply and compute using finite-element programs. Upon completion, the equivalent stress and strain of the RVE model are determined by

$$\bar{\varepsilon}_{ij} = \frac{1}{V} \int_V \varepsilon_{ij} dV \tag{3.6}$$

$$\bar{\sigma}_{ij} = \frac{1}{V} \int_V \sigma_{ij} dV \tag{3.7}$$

where V is the volume of the RVE model.

3.2.2   *Generalized Hill plasticity model*

The generalized Hill plasticity model is a framework for describing the anisotropic plastic behavior of materials [6,7]. It captures differences in yield strength under tension and compression, as well as the mechanical behavior during the plastic stage. The model incorporates the generalized Hill yield criterion, anisotropic hardening laws, and an associated flow rule. The expression for the yield surface of the generalized Hill yield criterion is given by

$$\{\sigma\}^{\mathrm{T}}[M]\{\sigma\} - \{\sigma\}^{\mathrm{T}}[L] - K = 0 \tag{3.8}$$

where $[M]$ is a matrix that encapsulates the variation of yield strengths across different directions, $[L]$ mean a matrix delineating the principal stress components, and K signifies the yield stress in a designated direction. Equation (3.8) is derived to obtain (3.9)

$$\begin{aligned}[M_{11}\sigma_{11}^2 + M_{22}\sigma_{22}^2 + M_{23}\sigma_{33}^2 + M_{44}\sigma_{12}^2 + M_{55}\sigma_{23}^2 + M_{66}\sigma_{31}^2 + 2M_{31}\sigma_{33}\sigma_{11} \\ + 2M_{23}\sigma_{22}\sigma_{33} + 2M_{31}\sigma_{33}\sigma_{11}] - [L_1\sigma_{11} + L_2\sigma_{22} + L_3\sigma_{33}] - \sigma_y = 0 \quad (3.9)\end{aligned}$$

where M_{jj} denotes a function pertaining to the yield stresses under tension and compression, defined as

$$M_{jj} = \frac{\sigma_y}{\sigma_{+j}\sigma_{-j}}, j = 1, 2, ..., 6 \tag{3.10}$$

where σ_y represents the stress in the y direction, parameter σ_{+j} represents the tensile yield strength in each direction, parameter σ_{-j} represents the compressive yield strength in each direction, with specific values determined based on the material properties. The subscript indicating tensile and compressive yield strengths is denoted by $\{1, 2, 3, 4, 5, 6\}$ signifying $\{11, 22, 33, 12, 23, 31\}$.

According to the incompressibility assumption in plastic mechanics, the mixed subscript coefficients are determined by

$$M_{12} = -\frac{1}{2}(M_{11} + M_{22} - M_{33}) \tag{3.11}$$

$$M_{12} = -\frac{1}{2}(M_{11} + M_{22} - M_{33}) \tag{3.12}$$

$$M_{23} = -\frac{1}{2}(M_{11} + M_{22} - M_{33}) \tag{3.13}$$

where L_j is the strength coefficient, defined as

$$L_j = M_{ii}(\sigma_{+j} - \sigma_{-j}), j = 1, 2, 3 \tag{3.14}$$

In the elastic phase, it is necessary to specify the material's elastic modulus and Poisson's ratio. Upon entering the plastic phase, the generalized Hill yield criterion requires 18 constants, including the yield strengths in various directions, shear modulus, and transverse modulus. These parameters are determined based on specific RVE models.

Additionally, the strengthening rule of the generalized Hill plastic model is an isotropic strengthening criterion used to determine the size of subsequent

yield surfaces, reflecting the history of plastic loading. The incremental constitutive relationship in the plastic phase is governed by the flow rule and
incremental constitutive law. The associated flow rule determines the magnitude of strain, while consistency conditions establish the direction of strain
increments, which corresponds to the outward normal direction of the yield
surface.

3.2.3 Analysis process of coupling RVE method with generalized Hill plastic model

The analysis process of the coupling algorithm based on the RVE method and the
generalized Hill plastic model is shown in Figure 3.2. First, a hierarchical multiscale analysis method is adopted to analyze the TSV wafer, and periodic boundary
conditions are applied to the selected RVE model. Tensile and shear experiments
are simulated in ABAQUS. Upon completion, according to the direct homogenization theory, (3.6) and (3.7) are utilized to obtain the equivalent stress and
strain of the RVE model at each sub-step. This allows for the generation of stress–
strain curves for the RVE model in different directions. From these curves, yield
stress, tangent modulus, and shear modulus are extracted.

Finally, the obtained parameters are inputted into an anisotropic homogeneous model with the same geometric dimensions as the RVE model. This
homogeneous model is then used to replace the RVE model, which represents
the actual model, thus completing the calculation of the coupling algorithm.
The generalized Hill plastic model is employed to describe the anisotropic
plastic behavior of the homogeneous model, with its parameters derived from
the RVE model. This approach enables a more accurate and efficient analysis of
the mechanical behavior of complex materials, incorporating both their elastic
and plastic properties.

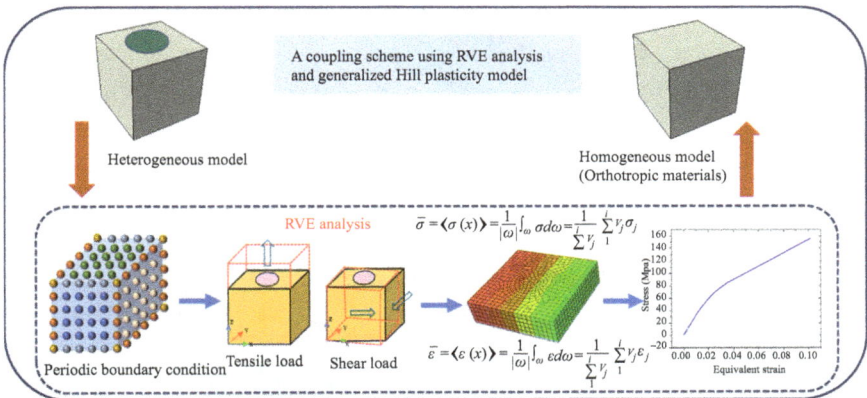

Figure 3.2 Analysis process of coupled scheme

3.3 Examples investigation and discussion

3.3.1 2.5D packaging structure

Figure 3.3 depicts the geometric model of a 2.5D packaging structure. At the bottom lies a PCB substrate with dimensions of 8 mm by 8 mm and a thickness of 0.8 mm. Gaps between solder balls are filled with underfill, reducing stress concentration and enhancing overall structural reliability. Beneath the chip, a micro-bump layer consists of periodically distributed copper bumps, sintered silver, and epoxy resin. The height of the bump layer is 150 μm, assuming circular cross-sections for copper and sintered silver bumps, with epoxy molding compounds (EMC) filling other areas of the solder balls. The bumps are spaced 0.2 mm apart. The periodic distribution of micro-bumps within the epoxy encapsulant includes three layers per micro-solder joint: upper and lower layers of copper bumps, with the upper layer connected to the chip's pad and the lower layer interconnected with the interposer. Sintered silver solder material in between facilitates signal and mechanical connections between copper bumps. Table 3.1 provides structural parameters for the substrate, silicon, underfill, EMC, and sintered silver solder material.

3.3.2 Establishment of the RVE model and acquisition of equivalent material parameters

Uniformization strategy involves the equivalency of the micro-bump layer, beginning with the establishment of the RVE model. Based on the structural characteristics of micro-bumps, a single micro-bump is considered to comprise two types of

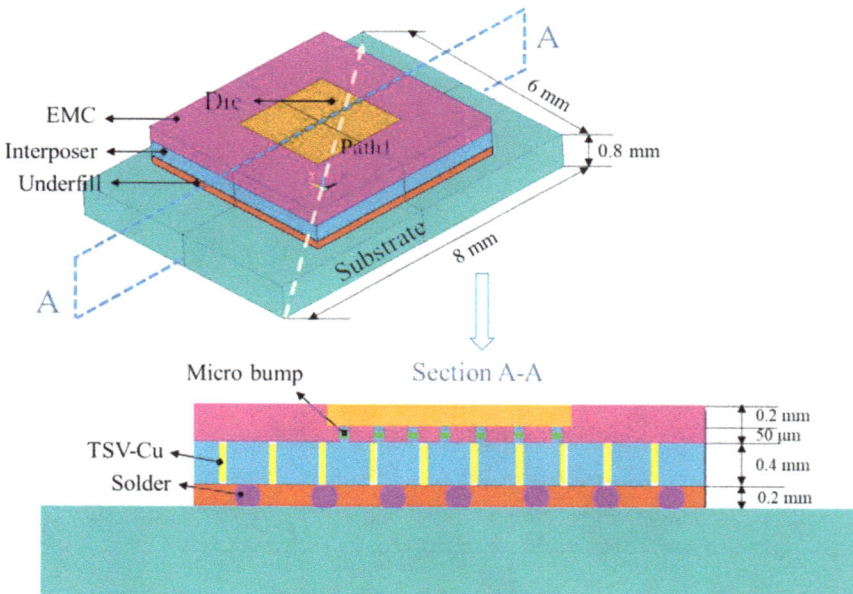

Figure 3.3   Geometry model of 2.5D packaging structure

Table 3.1 2.5D package structure size parameters

Components	Size (mm)
PCB substrate	8 × 6 × 0.8
Pitch between solder balls	0.6
Diameter of solder balls	0.25
Height of solder balls	0.2
Thickness of underfill	0.2
Size of interposer	5 × 5 × 0.4
Thickness of EMC	0.25
Chip size	0.40 × 0.40 × 0.05

RVE models. The local magnification of the micro-bump is depicted in Figure 3.3. One type consists of EMC and copper bumps, while the other includes EMC and sintered silver solder. Following the developed coupling algorithm, after determining the RVE model, symmetric mesh division and periodic boundary conditions are applied. Subsequently, in ABAQUS, periodic boundary conditions are enforced using multiple-point constraint equations

$$C_1 u_{p_1}^i + C_2 u_{p_2}^i + \cdots + C_n u_{p_n}^i = 0 \tag{3.15}$$

where C_n denotes the coefficient determined based on the positions of the node and the reference point, n represents the number of terms in the linear multi-point constraint equation (set to 3 in this study), u_p^i corresponds to the displacement variables, i indicates the direction of displacement (where $i = 1, 2, 3$), and p_n describes the position of node n.

After applying periodic boundary conditions, displacement loads are imposed using constraint equations. Based on simulation results, stress–strain curves in various directions under tensile, compressive, and shear loads can be obtained for the RVE model. Equivalent stresses and strains are determined using the expansions of (3.16) and (3.17). Ultimately, the constitutive behavior of the RVE model in different directions can be derived

$$\bar{\sigma} = \langle \sigma(x) \rangle = \frac{1}{|\omega|} \int_\omega \sigma d\omega = \frac{1}{\sum\limits_1^i V_j} \sum\limits_1^i V_j \sigma_j \tag{3.16}$$

$$\bar{\varepsilon} = \langle \varepsilon(x) \rangle = \frac{1}{|\omega|} \int_\omega \varepsilon d\omega = \frac{1}{\sum\limits_1^i V_j} \sum\limits_1^i V_j \varepsilon_j \tag{3.17}$$

where i represents the total number of elements in the RVE model, V_j denotes the volume of the jth element in the RVE model, σ_j corresponds to the stress of the jth element in the RVE model, and ε_j indicates the strain of the jth element in the RVE model.

To obtain the equivalent coefficient of thermal expansion (CTE) of the RVE model, predefined temperature fields are applied along with periodic boundary conditions as loads and boundary conditions. Strains on the nodes of the RVE model relative to the given temperature differential are extracted. The CTE in different directions is determined by defining $\alpha = \varepsilon/\Delta T$.

Establishing an RVE model with dimension $0.2 \times 0.2 \times 0.05$ mm is depicted in Figure 3.4. The dimensions of this model are the same as those of the RVE model composed of EMC and sintered silver. In ABAQUS, the RVE model is partitioned with symmetric meshing using C3D8R elements, with 3120 nodes and 2395 elements as specified in Table 3.2 for material parameters. To validate the effectiveness of applied boundary conditions, both phases of the RVE model are made from the same material. Therefore, the RVE model is modeled as a homogeneous solid block, and upon completion, unit strain $(E_{11}, E_{22}, E_{33}, E_{12}, E_{13})$ are applied to it. The obtained six uniform strain fields are shown in Figure 3.5. According to Li *et al.* [8], correct periodic boundary conditions were applied to the RVE model in this section.

After verifying the correctness of the applied periodic boundary conditions, the RVE model depicted in Figure 3.6 was subjected to tensile and shear deformations

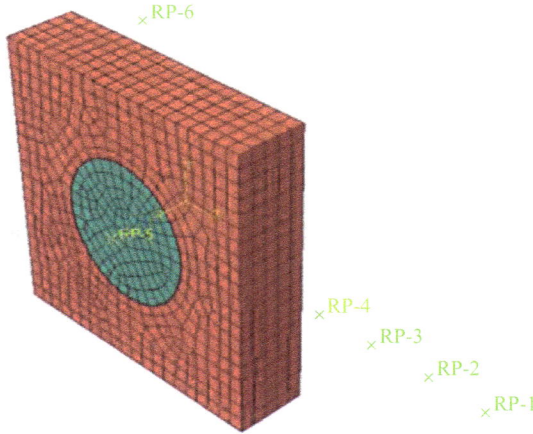

Figure 3.4   RVE model composed of EMC and sintered silver

Table 3.2   Material properties adopted in the simulations [9–12]

Materials	Elastic modulus (GPa)	Poisson's ratio	CTE (ppm/°C)
PCB	22	0.28	18.5
Sintered silver	27.76	0.22	24.5
Underfill	8.50	0.35	32
EMC	28.50	0.2	13
Copper	142.48	0.34	17
Silicon	130	0.28	3

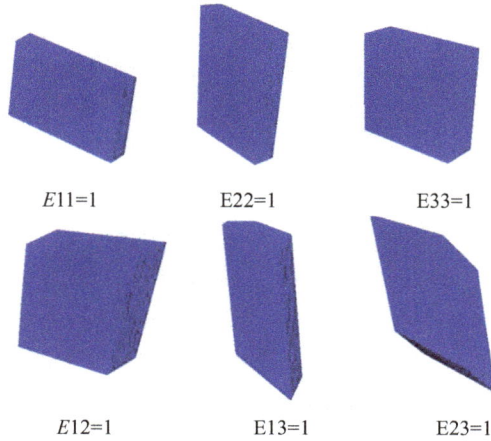

$E11=1$ $E22=1$ $E33=1$

$E12=1$ $E13=1$ $E23=1$

Figure 3.5 Homogeneous RVE model deformation under shear and tensile strain

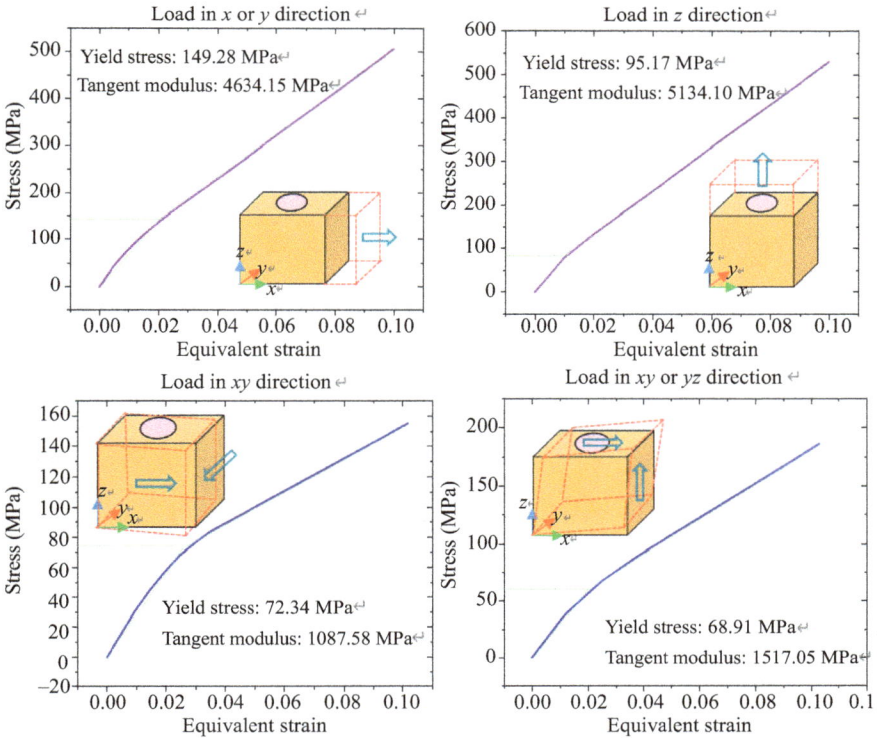

Figure 3.6 Averaged stress–strain relationship for the RVE model which consists of EMC and Cu bump

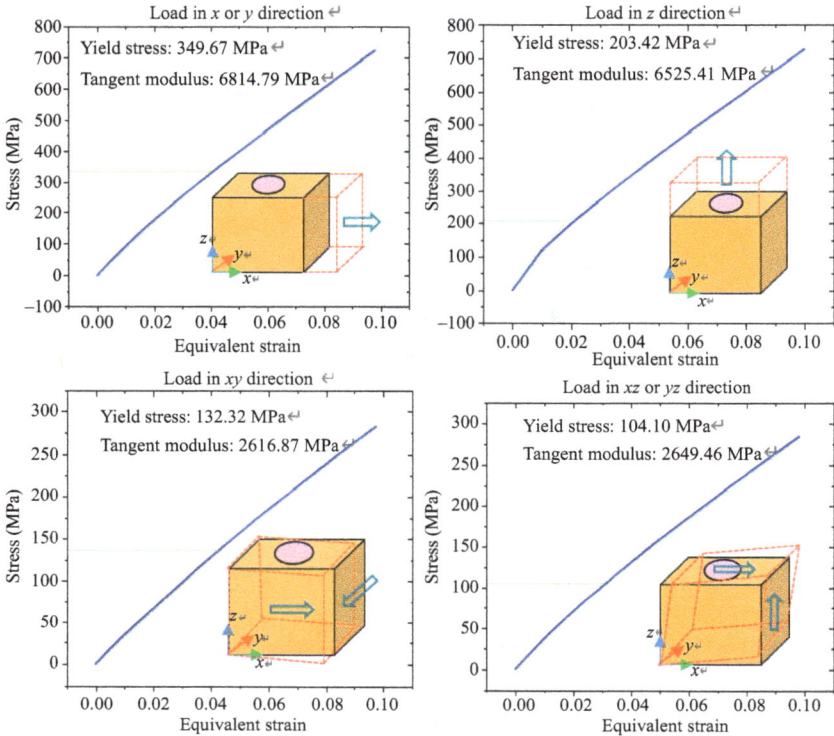

Figure 3.7   Averaged stress–strain relationship for the RVE model which consists of EMC and sintered sliver

using a coupled approach. Stress–strain curves in different directions were obtained accordingly. Based on the generalized Hill plasticity model, the stress–strain relationship during the plastic phase of the RVE was characterized using a bilinear elastic–plastic model.

Figure 3.6 shows the average stress–strain relationships of the RVE model composed of EMC and micro bumps, with annotations for yield stress, tangent modulus, and shear modulus under shear and tensile stresses in various directions. Figure 3.7 depicts the average stress–strain relationships of the RVE model composed of EMC and sintered silver, similarly fitted using a bilinear elastic model to obtain corresponding yield stress, tangent modulus, and shear modulus.

3.3.3   The establishment of finite-element models and application of boundary conditions

To validate the effectiveness of the proposed coupling method, two quarter finite-element models were established in this section. The first model, shown in Figure 3.8(a), is a detailed finite-element model comprising Solid185 elements totaling 575,397 units. The second model, depicted in Figure 3.8(b), is a

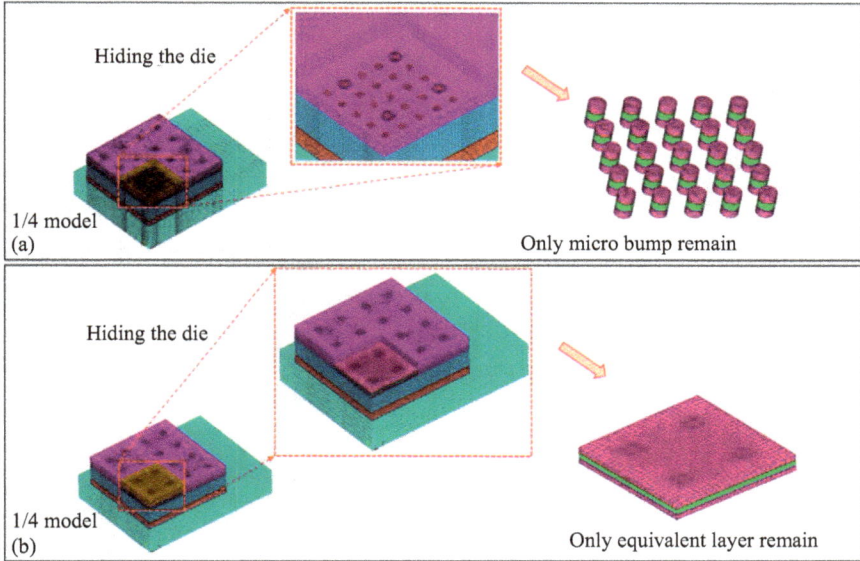

Figure 3.8   Quarter FE model of 2.5D packaging structure

homogenized equivalent finite-element model where the elements in the homogenized region are of type Solid45, while the rest remain Solid185, resulting in a total of 227,570 elements. In ANSYS, Solid185 and Solid45 are both eight-node three-dimensional solid elements; however, Solid45 elements allow for the use of a generalized Hill plasticity model, unlike Solid185 elements. The primary distinction between these two finite-element models lies in the uniformization of the micro-convex layer. Uniformizing the micro-convex layer beneath the chip mitigates the need for extensive mesh refinement, as this layer exerts the most significant influence on mesh partitioning. To focus solely on evaluating the proposed method's efficacy, structural elements such as RDL and Cu Pad were omitted from the numerical simulations to eliminate extraneous factors.

To ensure consistency between the two models, identical boundary conditions and loading conditions were applied. Symmetric boundary conditions were imposed on the YZ and XZ planes. Additionally, a central point on the substrate was fixed to constrain rigid body displacements of the entire model. Following the JEDEC thermal cycling standard [13], simulations were conducted to assess the solder ball's lifespan under thermal cycling conditions. Five temperature cycles were simulated in this section, each comprising five load steps per thermal cycle. The temperature varied from 25 °C to 125 °C and then cooled to −40 °C, resulting in a temperature range of 165 °C. The heating and cooling phases each lasted 10 min, with the high and low temperatures maintained for 5 min each. The temperature loading cycle is illustrated in Figure 3.9. Material parameters used in establishing the finite-element models are detailed in Tables 3.2 and 3.3.

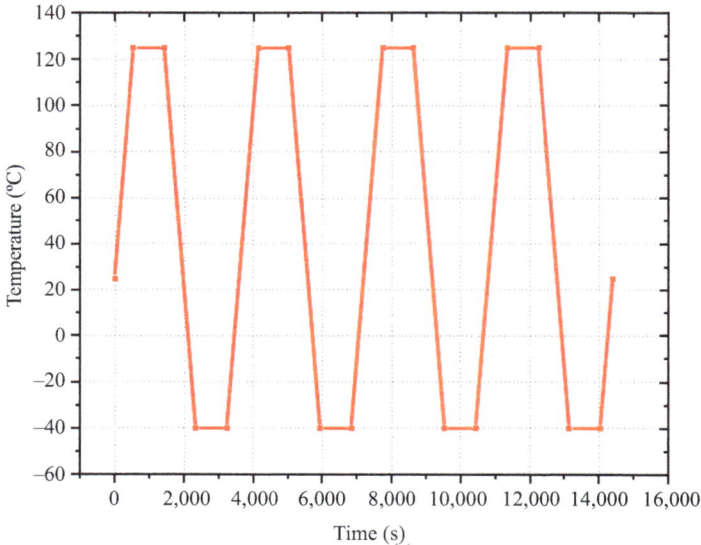

Figure 3.9   Simulation condition of temperature during thermal cycling

Table 3.3   Anand model parameter for sintered Ag [14]

Parameter	Symbol	Value
Pre-exponential factor	A (s^{-1})	9.81
Activation energy/universal gas constant	A/R (K)	5706
Multiplier of stress	ξ	11
Strain rate sensitivity of stress	m	0.66
Coefficient for deformation resistance saturation	s(MPa)	67.3
Deformation resistance	n	0.00326
Hardening/softening constant	$h_\circ$(MPa)	1.58×10^4
Strain rate sensitivity of hardening or softening	a	1
Initial deformation resistance	s_o(MPa)	2.77

3.4   Results and discussion

The calculated results of displacement, stress, strain, equivalent plastic strain, and fatigue life of solder joints obtained by the two models are compared in this section to verify the accuracy of the proposed method for predicting the fatigue life of solder joints. Figure 3.10 shows the comparison of the deflection variations along path 1 at the same time of 1445 s by the two models, where path 1 is indicated in Figure 3.3, along the diagonal direction of the substrate. It can be observed from the figure that the two curves match well. The two curves almost coincide at the middle position of the model, with the same deflection magnitude, while the deflection

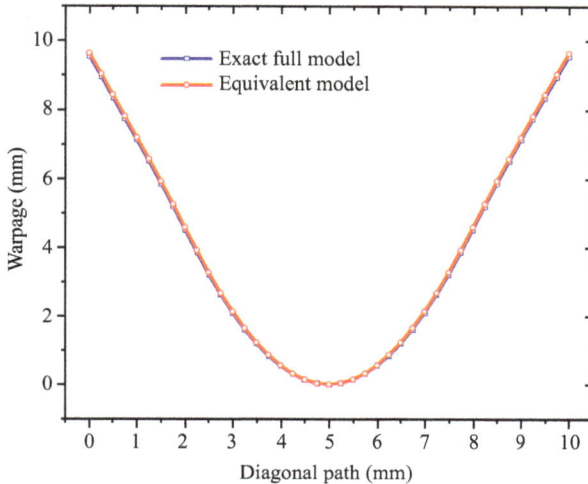

Figure 3.10   Comparison of warpage values of two models in the direction of path 1 at 1445 s

value becomes larger closer to the edges of the substrate. The maximum deflection value in the equivalent model is 9.6 µm, with an error of only 3% compared to the fine model.

Furthermore, at the 545th second of the thermal cycling curve, when the temperatures in both models are 125 °C, the Von mises stress, y-direction normal stress, and the first principal stress distributions on the surface of the chip are compared, as depicted in Figure 3.11. The comparison shows highly consistent distributions of these three stresses between the fine model and the equivalent model, with maximum stress values located in the top-right corner in both cases. This is attributed to severe thermal mismatch at the interface between the chip and the encapsulant, causing stress concentrations particularly at the corners. While there is some error in the maximum stress values at these corner stress concentrations, the computational results of both models are generally consistent in other non-stress concentrated areas.

In addition to static comparisons at specific time points, dynamic comparisons of the computed results were also conducted. Figure 3.12 illustrates the temporal variation of the first principal stress, von Mises stress, and normal stress at point A (as shown in Figure 3.13). Due to the applied thermal cyclic load, all three stress types exhibit periodic variations. The figure demonstrates that in both models, stresses increase or decrease non-linearly with temperature changes, showing periodic behavior over time. Within a cycle, the stress initially rises and, upon reaching the yield point, no longer increases. Comparison of the stress components from the equivalent model with those from the fine model reveals that the magnitude of each stress component in the equivalent model is consistent with that from the fine model, even during moments of rapid stress changes. These numerical

Exact full model   Equivalent model

7.28933
15.2437
23.1981
31.1525
39.1069
47.0613
55.0158
62.9702
70.9246
78.879

(a)

-32.16
-22.4783
-12.7965
-3.11474
6.56703
16.2488
25.9306
35.6123
45.2941
54.9758

(b)

-.182328
8.99259
18.1675
27.3424
36.5173
45.6923
54.8672
64.0421
73.217
82.3919

(c)

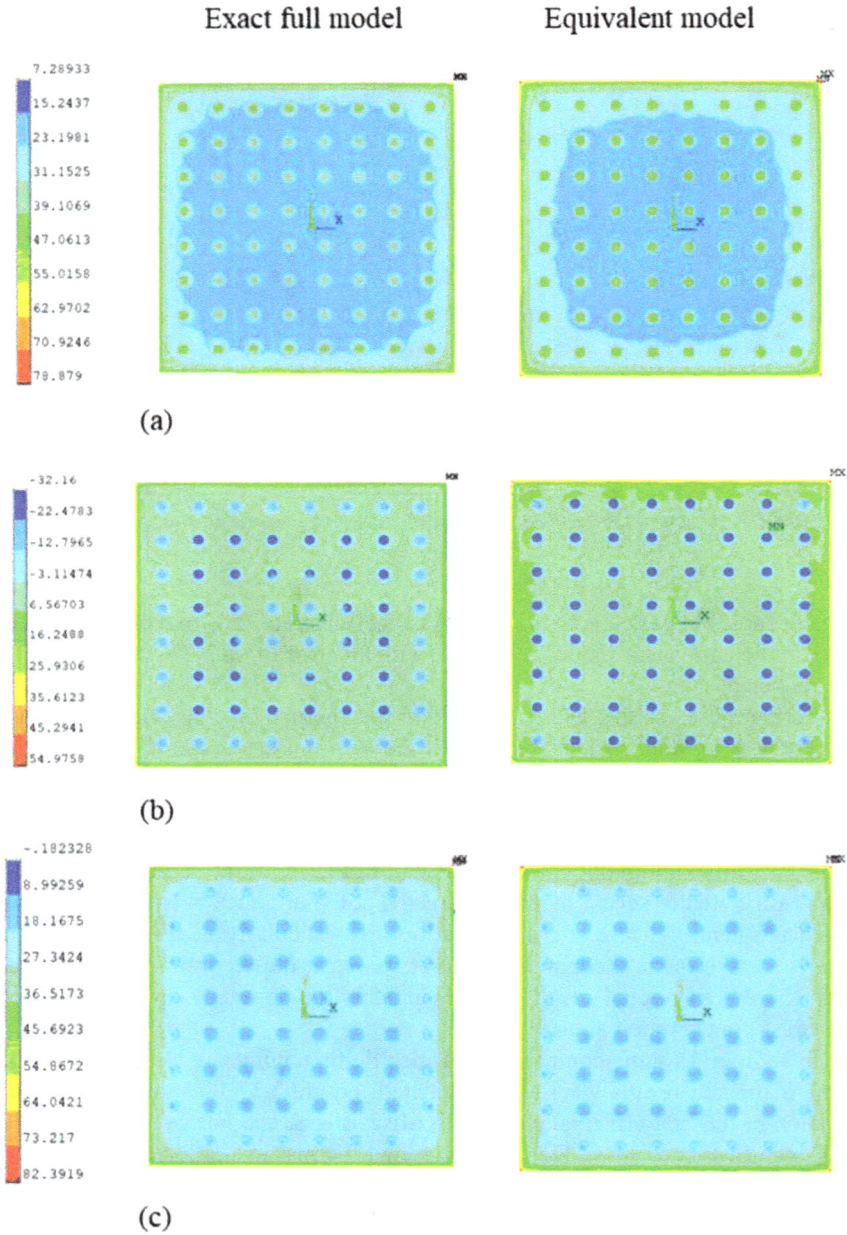

Figure 3.11   Stress contour at die top surface at 545 s (a) von Mises stress
(b) Normal stress at y direction (c) First principle stress

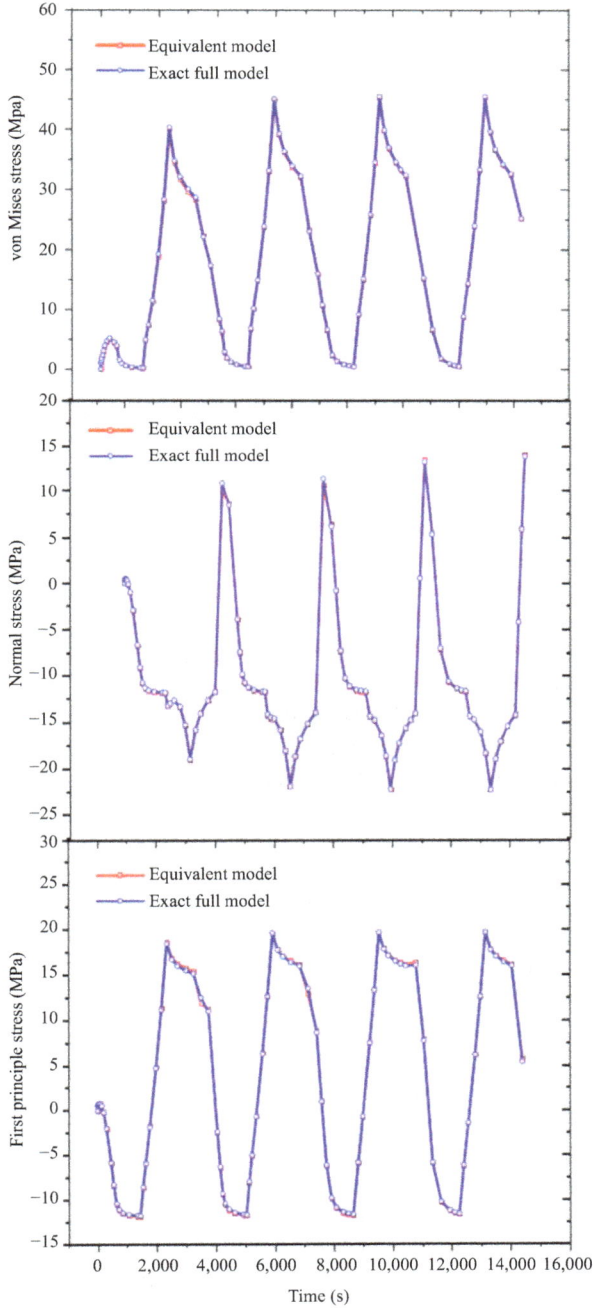

Figure 3.12   Comparison of the stress evolution between the exact full model and equivalent model.

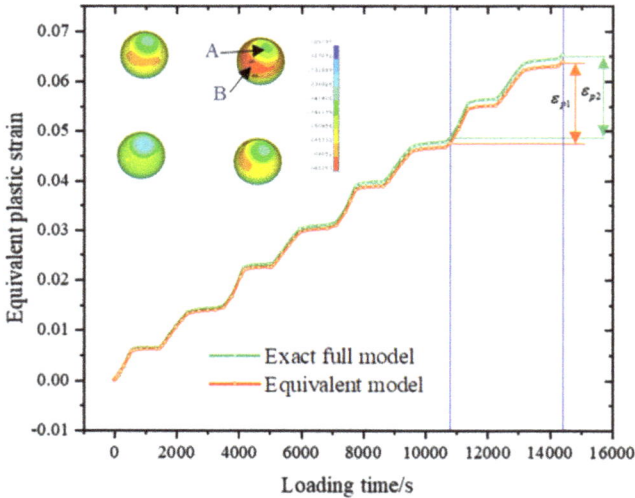

Figure 3.13   Equivalent plastic strain in sintered sliver solder in thermal cycling load

results indicate that the equivalent model maintains a high level of computational accuracy.

Figure 3.13 depicts the temporal evolution of equivalent plastic strain at critical point B under thermal cyclic loading for both models. It is observed from the figure that the equivalent plastic strain obtained from both models increases continuously in a step-like manner over time. This increase occurs during temperature rise or fall stages, while the equivalent plastic strain remains stable during the dwell periods. Since equivalent plastic strain accumulates over plastic deformation processes, its value only increases and does not decrease. Therefore, whether heating or cooling, equivalent plastic strain will increase. Additionally, based on the strain increment at the critical point during the final cycle, the fatigue life of solder balls is determined using the Coffin–Manson formula. The thermal fatigue life of solder balls computed by the equivalent model is 2134 cycles, with an error of 0.27% compared to the comprehensive model results.

Table 3.4 compares various aspects of the computational process, including the amount of memory required, the number of elements and nodes, and the central CPU computation time for both models. The detailed model has 594,588 nodes and 575,397 elements. In contrast, the equivalent model's nodes and elements are reduced by 60.11% and 60.45%, respectively. Both models utilize the same computer configuration (Intel Xeon E5-2696 V4 central processor and 64 GB of memory). The detailed model's numerical analysis requires 389 min to complete, while the equivalent model requires only 97 min. The comparison results indicate that the equivalent model achieves the same level of computational accuracy as the detailed model, but the uniformization of the real model to establish the equivalent

Table 3.4   Comparison of computational efficiency between two models

	Exact full model	Equivalent model
Element type	Solid185	Solid185, Solid45
The number of elements	575,397	227,570
The number of nodes	594,588	237,176
CPU computation time	389 min	97 min
Storage space	91.8 GB	36.1 GB
Equivalent plastic strain range at the last cycle	0.01634	0.01673
Fatigue life	2193	2134

model significantly reduces the required computation time and resources compared to the detailed model, highlighting the advantages of the coupled algorithm. Therefore, the coupled method proposed in this chapter can be effectively applied to the micro solder joints of 2.5D packaging structures for equivalent modeling and accurately predicting fatigue life.

3.5   Chapter summary

This chapter introduces a novel coupled analysis method that leverages the RVE technique in conjunction with the generalized Hill plasticity model. This coupling algorithm is specifically designed for predicting the thermal fatigue life of solder joints in 2.5D packaging structures, a critical aspect in the reliability of modern electronic systems. The primary innovation of this approach is its ability to extract plasticity parameters from the homogenized region of micro solder joints and integrate them into a simplified homogeneous model, effectively replacing the detailed, computationally expensive model typically used in such simulations. By doing so, this method reduces both the computational time and memory storage requirements without sacrificing significant accuracy in predicting the thermal fatigue life.

The effectiveness of the coupling algorithm is validated through a comparative analysis between the detailed model and the equivalent model under thermal cyclic loading conditions. One of the key findings from this analysis is that, at 545 s into the thermal cycle, the maximum deflection error between the two models is a mere 3%. This suggests that the equivalent model can reliably replicate the behavior of the detailed model with a high degree of precision. Furthermore, the stress distribution on the surface of the chip, as obtained from both models, is consistent, with the location of maximum stress being identical in both cases. This provides further validation for the equivalence of the two models in terms of predicting critical stress concentrations, which are crucial in assessing the risk of failure in solder joints.

The temporal evolution of stress at point A in the models shows that the stress amplitude and trend closely match between the detailed and equivalent models, reinforcing the accuracy of the proposed coupling algorithm. Additionally, when identical thermal cyclic loading is applied to both models, the temporal evolution of equivalent plastic strain at the critical point of the solder ball is extracted. The comparative analysis of these curves reveals a strong correlation between the results from both models, demonstrating that the equivalent model can effectively capture the complex plastic deformation behavior that occurs in the solder joints during thermal cycling.

Using the Coffin–Manson formula, the predicted fatigue life of the solder ball is calculated based on the results from both models. The life prediction based on the equivalent model deviates from that of the detailed model by only 0.27%, which further emphasizes the high accuracy of the equivalent model in predicting the thermal fatigue life. Despite the remarkable accuracy, the equivalent model offers significant computational advantages. It reduces the overall computation time by approximately 75% compared to the detailed model and requires only 39.89% of the storage space. This substantial reduction in computational resources makes the equivalent model a highly efficient tool for simulating thermal fatigue life in practical applications, especially when dealing with large-scale 2.5D packaging systems.

The comparison of computational results between the detailed and equivalent models clearly demonstrates that the proposed coupling algorithm is not only accurate but also efficient. The homogenization of critical structures, such as the periodically distributed micro-protrusions in the solder joints, is accomplished in a rational and effective manner. The high accuracy of the finite-element models generated using this method confirms its suitability for predicting the performance and durability of solder joints in complex 2.5D packaging structures. As a result, the developed coupling algorithm offers a promising solution to expedite the numerical simulation process, significantly reducing both computational time and resource consumption while maintaining a high level of accuracy.

In conclusion, the coupling algorithm introduced in this chapter represents a significant advancement in the field of electronic packaging reliability analysis. By providing an efficient and accurate method for simulating thermal fatigue life in solder joints, this approach holds great potential for accelerating the design and optimization of 2.5D packaging structures. Furthermore, the reduction in computational costs makes it feasible to perform extensive simulations across a variety of design configurations, thereby facilitating the development of more reliable and durable electronic systems in the future.

References

[1] H. Kim, J.Y. Hwang, S.E. Kim, Y.-C. Joo, and H. Jang, Thermomechanical challenges of 2.5-D packaging: A review of warpage and interconnect reliability, *IEEE Transactions on Components, Packaging and Manufacturing Technology Information* 13 (2023) 1624–1641.

[2] C. Zhang, X.W. Xu, and S.X. Guo, Microstructure model and finite element analysis of mechanical properties of 2D 1×1 biaxial braided composites, *Acta Materiae Compositae Sinica* 28 (2011) 215–222.

[3] C. Zhang, X.W. Xu, and X.J. Xu, Research progress in finite element analysis on macro-meso mechanical properties of 3D multi-directional braided composites, *Acta Materiae Compositae Sinica* 32 (2015) 1241–1251.

[4] P.M. Suquet, Introduction, in: E. Sanchez-Palencia, A. Zaoui (eds.) , *Homogenization techniques for composite media*, Springer, Berlin, 1987: pp. 193–198.

[5] Z. Xia, Y. Zhang, and F. Ellyin, A unified periodical boundary conditions for representative volume elements of composites and applications, *International Journal of Solids and Structures* 40 (2003) 1907–1921.

[6] C.F. Shih and D. Lee, Further developments in anisotropic plasticity, *Journal of Engineering Materials and Technology* 100 (1978) 294–302.

[7] S. Valliappan, P. Boonlaulohr, and I.K. Lee, Non-linear analysis for anisotropic materials, *Numerical Meth Engineering* 10 (1976) 597–606.

[8] S. Li, C. Zhou, H. Yu, and L. Li, Formulation of a unit cell of a reduced size for plain weave textile composites, *Computational Materials Science* 50 (2011) 1770–1780.

[9] K. Wakamoto, Y. Mochizuki, T. Otsuka, K. Nakahara, and T. Namazu, Temperature dependence on tensile mechanical properties of sintered silver film, *Materials* 13 (2020) 4061.

[10] Y.D. Li, P. Chen, F. Qin, *et al.*, Constitutive modelling of annealing behavior in through silicon vias-copper, *Materials Characterization* 179 (2021), 111359.

[11] G.N. Yang, Z.L. Kuang, H.Q. Lai, *et al.*, A quantitative model to understand the effect of gravity on the warpage of fan-out panel-level packaging, *IEEE Transactions on Components, Packaging and Manufacturing Technology* 11 (2021), 2022–2030.

[12] Z. Chen, Z. Zhang, F. Dong, S. Liu, and L. Liu, A hybrid finite element modeling: artificial neural network approach for predicting solder joint fatigue life in wafer-level chip scale packages, *Journal of Electronic Packaging* 143 (2021), 011001.

[13] JEDEC Solid State Technology Association Announces 50th Anniversary, JEDEC.

[14] D. Yu, X. Chen, G. Chen, G. Lu, and Z. Wang, Applying Anand model to low-temperature sintered nanoscale silver paste chip attachment, *Materials & Design* 30 (2009) 4574–4579.

Chapter 4
Molding process simulation of substrates

4.1 Introduction

The encapsulation process of epoxy resin, which is commonly used in molded packaging, involves complex interactions among several critical factors, including the resin's flow characteristics (such as viscosity), the structural deformation of the mold and packaging materials, and the polymerization curing reactions of the resin. Traditionally, mold designs for this process relied heavily on trial-and-error methods to assess the reliability of transferring molded packaging. While this approach may have provided some insight, it was extremely time-consuming and resource-intensive. It often led to delays and unnecessary costs in the development cycle, especially when unexpected defects or inefficiencies were encountered during production.

In response to these challenges, the industry has increasingly turned to finite-element simulation methods for preemptive evaluation of process parameters, offering a more efficient and accurate means of predicting and optimizing outcomes before practical implementation. By simulating mold flow and curing behavior in advance, manufacturers can identify potential issues in the encapsulation process without the need for costly physical trials. This method facilitates a more controlled and informed approach to design, significantly reducing the need for trial-and-error adjustments.

Mold flow analysis using finite-element simulations plays a pivotal role in the mature and stable development of the encapsulation process [1,2]. It helps to optimize key parameters such as resin flow patterns, curing times, and cooling rates, all of which are crucial for ensuring a high-quality end product. In injection molding, achieving packaging reliability is not solely dependent on the final product design but also on a combination of other factors, including proper mold design, the material characteristics of the resin, and the selection of suitable processing parameters. Each of these elements must be carefully integrated to minimize the risk of defects and improve overall product quality.

One of the key challenges in transfer molding processes is ensuring that the structural design of the encapsulation aligns well with the mold's design and the process parameters used during production [3,4]. If these factors are not appropriately matched, various defects may arise, many of which are random or stochastic in nature. These defects can affect the functionality and reliability of the final product, especially when dealing with delicate components such as inductors on substrates, which may be prone to damage from warping or uneven resin distribution.

This chapter focuses on the use of mold flow analysis software to investigate critical factors such as resin conversion rates and warping behaviors under different process parameters in transfer molding. By simulating the effects of varying parameters such as injection pressure, mold temperature, and resin viscosity, the analysis aims to predict and optimize the performance of the encapsulation process. This approach not only helps to assess the feasibility of the encapsulation process before physical production begins, but it also enables the identification of conditions that may lead to excessive warpage or other structural issues. By addressing these issues proactively, the simulation helps to ensure that the encapsulated components, such as inductors, remain protected and maintain their integrity throughout the molding process.

In summary, by leveraging advanced mold flow analysis techniques, manufacturers can gain a deeper understanding of the complex interactions within the encapsulation process and make more informed decisions regarding process parameters and mold design. This leads to more efficient, cost-effective, and reliable production processes, minimizing defects and improving the overall quality of the final encapsulated components. The use of simulation tools is therefore essential in streamlining the encapsulation process, ensuring that products meet the necessary performance standards while minimizing waste and production time.

4.2   Curing process and cure-kinetic model

During the encapsulation and post-encapsulation curing processes in epoxy resin encapsulants, complex chemical cross-linking reactions occur between the epoxy resin and curing agent, commonly referred to as curing reactions. These reactions involve the gradual formation of large molecular networks from discrete monomers, accompanied by the release of substantial heat, as illustrated in Figure 4.1. Prior to the gelation point, the encapsulant behaves as a viscous fluid with nearly zero elastic modulus. Upon reaching the gelation point, significant changes occur in the material's mechanical behavior, transitioning it into a solid state with a markedly increased elastic modulus. As the curing reaction progresses, thermosetting polymers gradually transform from a viscous liquid state (region I) to a rubbery state (region II) and ultimately into a glassy state (region III).

As mentioned above, the curing reaction of epoxy resin encapsulants significantly influences their material characteristics and mechanical properties. The degree of curing of the encapsulant is a crucial parameter that reflects the extent of the curing reaction, closely associated with temperature and time. As the curing reaction progresses, the degree of curing gradually increases. Upon reaching the gelation point, the encapsulant transitions from a liquid to a solid state, undergoing hardening. Therefore, conducting rational experimental studies on the curing reaction of encapsulants is essential for understanding the curing mechanism, optimizing packaging process parameters, and enhancing manufacturability and

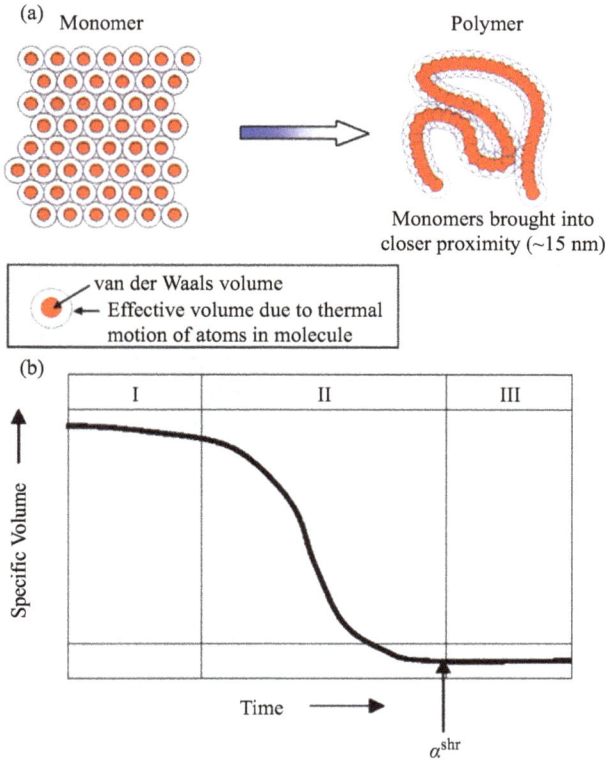

Figure 4.1   *Curing process and modulus evolution of EMC: (a) schematic of unit cell contraction occurring during the curing process and (b) three distinct regions in which physical properties of resin change throughout the polymerization process [5]*

reliability of encapsulation. In the field of electronic packaging, differential scanning calorimetry (DSC) is widely employed to investigate the curing reaction of encapsulants. DSC measures the rate of heat flow (exothermic or endothermic) relative to a reference material as a function of temperature under controlled temperature programs, thereby determining parameters such as curing heat release and relative degree of curing reaction [6].

Using the DSC method to determine the kinetic parameters of a curing reaction is based on two fundamental assumptions: first, it assumes that the heat generated from the curing reaction is directly proportional to the amount of reacted reactants; second, it assumes that the change in heat flow is directly proportional to the change in the degree of curing reaction. These assumptions are, respectively, represented by

$$\alpha = \frac{\Delta H_j}{\Delta H} \tag{4.1}$$

$$\frac{d\alpha}{dt} = \frac{1}{\Delta H}\frac{\Delta H_j}{dt} \tag{4.2}$$

where α represents the degree of curing, ΔH_j denotes the heat released by the curing reaction, and ΔH signifies the total heat of the curing reaction. The total heat of the curing reaction ΔH_{ult} is formulated as

$$\Delta H_{ult} = \Delta H_{iso} + \Delta H_{re} \tag{4.3}$$

where ΔH_{iso} represents the heat released during isothermal curing and ΔH_{res} denotes the dynamic residual heat after isothermal curing.

There are two methods for measuring heat flow changes using DSC to obtain curing kinetics models: isothermal and non-isothermal (or dynamic). For instance, Tamil *et al.* [7] conducted experimental studies on the curing reaction of encapsulants using both isothermal and dynamic DSC methods, obtaining kinetic model curves for the encapsulant under both conditions. Typically, several models are used to describe the curing kinetics of encapsulants and other thermal-setting polymer materials.

1. *N*th order kinetic model

$$\frac{d\alpha}{dt} = k(1-\alpha)^n \tag{4.4}$$

2. Autocatalytic kinetic model

$$\frac{d\alpha}{dt} = k\alpha^m(1-\alpha)^n \tag{4.5}$$

3. Non-autocatalytic and autocatalytic composite reaction models

$$\frac{d\alpha}{dt} = k_1(1-\alpha)^p + k_2\alpha^m(1-\alpha)^n \tag{4.6}$$

4. *N*th order kinetic model combined kinetic model

$$\frac{d\alpha}{dt} = (k_1 + k_2\alpha^m)(1-\alpha)^n \tag{4.7}$$

where α represents the degree of cure, m, n, p denote the reaction order, and k, k_1, k_2 employs the Arrhenius relationship dependent on temperature, as shown in the following equation:

$$k_i = A_i \exp\left(-\frac{E_i}{RT}\right) \tag{4.8}$$

where A_i represents the pre-exponential factor, E_i denotes the activation energy of the curing reaction, i is a subscript taking values 1 and 2, R stands for the Boltzmann gas constant, and T represents absolute temperature.

In practice, among the curing kinetics models mentioned above, the most suitable model for describing the curing reaction process of thermosetting polymer materials such as molding compounds is the combined *n*-order and autocatalytic composite reaction model, also known as the Kamal model [6,8].

4.3   Viscoelastic constitutive model

In electronic packaging materials, molding compounds are typical viscoelastic materials, whose mechanical properties depend not only on time but also on temperature. During the curing reaction process, molding compounds form large molecular structures from discrete monomers, and their degree of cure is closely related to process parameters such as time and temperature. The degree of cure of molding compounds also significantly affects their mechanical properties. Therefore, in numerical simulations of packaging processes, selecting an appropriate constitutive model for molding compounds is crucial for accurately predicting and enhancing the manufacturability and reliability of packaging. Kim *et al.* [9] found that compared to linear elastic constitutive models, viscoelastic constitutive models better predict the warpage of packages. Similarly, De Vreugd *et al.* [10] and Luan [11] concluded that finite-element results using viscoelastic constitutive models are more consistent with experimental measurements. Cho and Jeon [12] studied the effect of elastic modulus of molding compounds at different degrees of cure on the warpage of thin small outline packages (TSOPs), using a linear elastic constitutive model for molding compounds, revealing significant differences in finite-element results at different degrees of cure. Chiu *et al.* [13,14] investigated the warpage of molding compound/metallic laminate and chip-scale packages (CSP) during the post-molding curing process, employing a curing-related viscoelastic constitutive model considering curing shrinkage for molding compounds, and found excellent agreement between finite-element results and experimental measurements. These studies underscore the critical importance in numerical simulations of establishing constitutive models that accurately describe the mechanical properties of molding compounds for correctly predicting and enhancing the manufacturability and reliability of packaging.

For viscoelastic materials like molding compounds, their mechanical properties exhibit characteristics of both elastic solids and viscous fluids. The viscoelastic constitutive model is formulated as follows:

$$\sigma_i(t) = \int_{-\infty}^{t} C_{ij}(t - \zeta, T) \cdot \left(\frac{d\varepsilon_j(\zeta)}{d\zeta} - \frac{d\varepsilon_j^*(\zeta)}{d\zeta} \right) d\zeta \tag{4.9}$$

where C_{ij} represents the relaxation modulus, ξ denotes the duration of applied load, T signifies temperature, and ε_j^* encompasses initial strains, including those induced by temperature and curing. For isotropic materials, the relaxation modulus C_{ij} is composed of two mutually independent parameters, namely the shear relaxation modulus $G(t, T)$ and the bulk relaxation modulus $K(t, T)$, as follows:

$$C_{ij}(t, T) = K(t, T)V_{ij} + G(t, T)D_{ij} \tag{4.10}$$

where V_{ij} and D_{ij} represent constants of the volumetric and offset matrix, respectively. Thus, research on viscoelastic constitutive models essentially focuses on studying the shear relaxation modulus $G(t, T)$ and the bulk relaxation modulus $K(t, T)$ within the constitutive model.

The viscoelastic response of molding compounds is described by the time–temperature superposition principle, where the influence of temperature (time) on the relaxation modulus can be equivalently expressed through time (temperature). This relationship is depicted in (4.11), where the relaxation modulus can represent either the shear relaxation modulus or the bulk relaxation modulus

$$E(T_0, t) = E(T_1, t/a_T) \tag{4.11}$$

where T_0 and T_1 represent the temperature, t denotes the time, and a_T is the time–temperature shift factor. The time–temperature shift factor a_T is obtained directly through numerical methods from experimental data. The primary descriptive methods for this factor include the Williams–Landel–Ferry (WLF) equation, the Arrhenius equation, and the Vogel–Fulcher–Tamman (VFT) equation.

WLF equation:

$$\log a_T = \frac{-C(T - T_{ref})}{C_2 + T - T_{ref}} \tag{4.12}$$

where T_{ref} denotes the reference temperature and C_1 and C_2 are constants.

Arrhenius equation:

$$\log a_T = \frac{-\Delta E}{2.30R} \left(\frac{1}{T} - \frac{1}{T_{ref}} \right) \tag{4.13}$$

where E represents the activation energy, while R denotes the Boltzmann gas constant.

Vogel equation:

$$\log a_T = \log \frac{\tau_n(T)}{\tau_n(T_{ref})} = \frac{C}{T - T_\infty} - \frac{C}{T_{ref} - T_\infty} \tag{4.14}$$

where τ_n represents the relaxation time, C is a constant, and T_∞ is the temperature constant.

The shear relaxation modulus and bulk relaxation modulus can be represented by the generalized Maxwell equation in the form of a Prony series

$$G(t, T) = G_e + \sum_{n=1}^{N} G_n \exp\left(-\frac{t}{\tau_n} \right) \tag{4.15}$$

$$K(t, T) = K + \sum_{n=1}^{N} K_n \exp\left(-\frac{t}{\tau_n} \right) \tag{4.16}$$

where G_e and K_e, respectively, denote the equilibrium shear relaxation modulus and equilibrium bulk relaxation modulus, G_n and K_n represent the shear modulus and volume modulus of the nth Maxwell unit, n denotes the relaxation time, and N signifies the number of Maxwell units.

The above description pertains to fully cured encapsulant materials, where the viscoelastic mechanical properties of the encapsulant are independent of the curing reaction. However, during the encapsulation process and subsequent curing process, the encapsulant undergoes curing reactions, and the viscoelastic mechanical properties of the material are closely related to the degree of curing. In other words, the curing reaction can affect the viscoelastic behavior of the material. With the increase in the degree of curing, the material transitions from a viscous liquid to a viscoelastic solid, leading to significant changes in the relaxation modulus. There are three main aspects of changes: first, an increase in the modulus of the glassy state; second, an extension of the relaxation time; and finally, an increase in the equilibrium relaxation modulus, corresponding to the rubbery state modulus. The viscoelastic constitutive model related to curing is illustrated as

$$\sigma_i(t) = \int_{-\infty}^{t} C_{ij}(\alpha, t - \zeta, T) \cdot \left(\frac{d\varepsilon_j(\xi)}{d\xi} - \frac{d\varepsilon_j^*(\xi)}{d\xi} \right) d\xi \tag{4.17}$$

where $i, j = 1, ..., 6$, α is the degree of curing. The shear relaxation modulus and volume relaxation modulus related to curing are denoted as

$$G(\alpha, t, T) = G_e(\alpha) + \sum_{n1}^{N} G_n(\alpha, t) \exp\left(-\frac{t}{\tau_n(\alpha, t)} \right) \tag{4.18}$$

$$K(\alpha, t, T) = K_e(\alpha) + \sum_{n1}^{N} K_n(\alpha, t) \exp\left(-\frac{t}{\tau_n(\alpha, t)} \right) \tag{4.19}$$

where $G_e(\alpha)$ and $K_e(\alpha)$ represent the equilibrium shear modulus and equilibrium bulk modulus, respectively.

Adolf and Chambers [15] and Adolf and Martin [16] found in their study on the curing-related mechanical behavior of thermosetting polymer materials that the equilibrium shear modulus is a function of the degree of curing. When the curing reaction reaches the gelation point, the equilibrium shear modulus gradually increases from zero, establishing the Adolf model as

$$G_e(\alpha) = G_e^{\infty} \left[\frac{\alpha^2 - \alpha_{gel}^2}{1 - \alpha_{gel}^2} \right]^{8/3} \tag{4.20}$$

where G_e^{∞} denotes the equilibrium shear modulus at complete curing and α_{gel} represents the degree of curing at the gelation point.

For the equilibrium bulk modulus, Yang *et al.* [17] employed a method similar to the Adolf model to establish its relationship with the degree of curing, as follows:

$$K_e(\alpha) = K_1 + (K_E^{\infty} - K_1) \left[\frac{\alpha^2 - \alpha_{gel}^2}{1 - \alpha_{gel}^2} \right]^{8/3} \tag{4.21}$$

where K_e^∞ represents the equilibrium bulk modulus and K_1 denotes the volume modulus in the liquid state.

The curing-related relaxation time $\tau_n(\alpha, T)$ can be expressed as follows:

$$\tau_n(\alpha, T) = \tau_n^{ref} \cdot a_T \cdot a_\alpha \tag{4.22}$$

where τ_n^{ref} denotes the relaxation time at a reference temperature, curing degree a_T represents the time–temperature shift factor, a_α represents the time–curing degree shift factor. The time–curing degree shift factor a_α can be formulated using equations akin to the Arrhenius equation (4.13) and Vogel equation (4.14). Adolf *et al.* [18] initially explored the curing-related mechanical behaviors of thermosetting polymers based on the time–curing degree superposition principle, while subsequent studies by other scholars have largely built upon Adolf *et al.*'s foundational work. Simon *et al.* [19], by considering between T_g temperature and curing degree within the time–temperature shift factor a_T, have modified the Arrhenius equation and the Vogel equation.

References [20,21] omitted the viscoelastic properties of the material, asserting that the elastic modulus is solely dependent on the curing degree. Common models are represented as

$$E = (1 - \alpha_{mod})E_m^0 + \alpha_{mod}E_m^\infty + \gamma\alpha_{mod}(1 - \alpha_{mod})(E_m^\infty - E_m^0) \tag{4.23}$$

where $\alpha_{mod} = \frac{\alpha - \alpha_{gel}^{mod}}{\alpha_{diff}^{mod} - \alpha_{gel}^{mod}}$; $-1 < \gamma < 1$, α_{gel}^{mod} represents the curing degree at the gel point, α_{diff}^{mod} represents the curing degree at complete curing, E_m^0 denotes the elastic modulus at the uncured state, E_m^∞ denotes the elastic modulus at complete curing, and γ represents the coefficient describing the competitive mechanisms of stress relaxation and curing hardening effects.

4.4   Warpage evolution of substrate during the transfer molding process

4.4.1   Establishment of the FE model

Figure 4.2 illustrates a solid model constructed using 3D software Rhino, with the right half of the model encapsulated. The dimensions of the substrate are 180 mm × 90 mm × 0.45 mm, while those of the encapsulation are 78 mm × 40 mm × 1.2 mm. The interface dimensions of the runners and gates are 2 mm × 1.2 mm. To reduce finite-element computation time, the inductance model was simplified; the internal and external electrodes of the inductor were neglected, with dimensions of 0.4 mm × 0.2 mm × 0.36 mm, arranged in a 12 × 6 array staggered across the substrate. The vertical spacing is 6.7 mm, and the horizontal spacing is 6.4 mm.

The meshing of the model was carried out using the software Rhine and the plugin of Moldex3D. A total of 478,158 solid meshes were generated, with refined meshing at the location of the inductor, exhibiting good mesh quality, as shown in Figure 4.3.

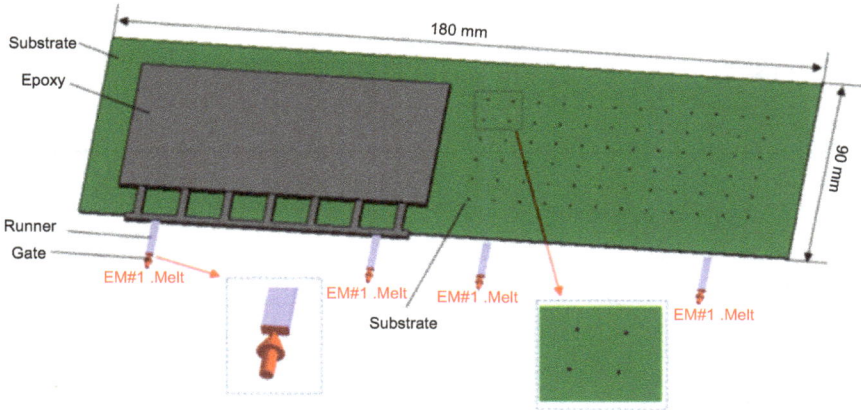

Figure 4.2   The packaging structure

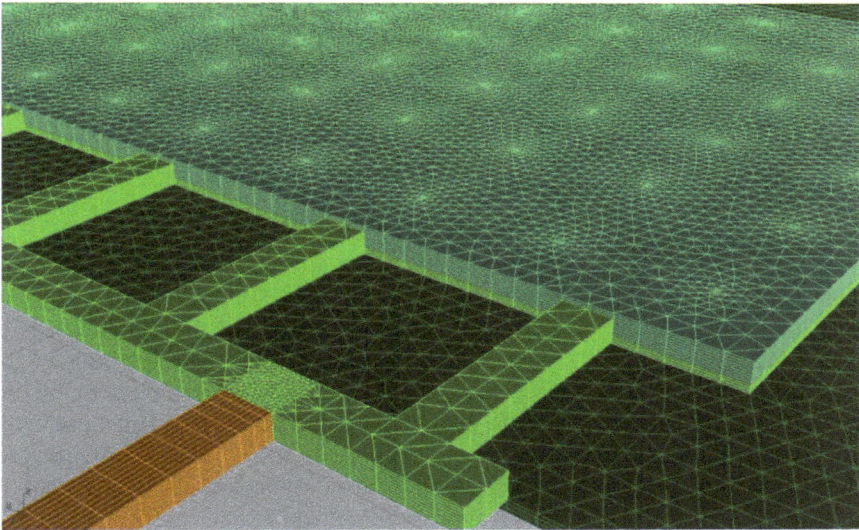

Figure 4.3   The mesh of the packaging structure

Finite-element simulation and analysis of transfer molding encapsulation were conducted using Moldex3D software. Prior to simulating the transfer molding encapsulation, it is essential to determine the mechanical performance parameters of the materials used, as the accuracy of these parameters directly impacts the finite-element analysis results. For this purpose, viscosity model parameters were sourced from Table 4.1, while the curing reaction kinetics and viscoelastic models were derived from experimental testing. Additional material mechanical performance parameters are provided in Table 4.2, including elastic modulus (E),

Table 4.1   Numerical parameters for the Cross-Castro–Macosko model [22]

n	τ^* (Dyne/c)	B	$T_b(K)$	C_1	C_2	a_g
0.293445	1000	1.55155e−2	22558.5	0.299842	12.7742	0.2

Table 4.2   Material parameters

Materials	E (GPa)	v	CTE (10^{-6}/°C)	K (W/m·K)
EMC	18	0.4	34	0.970
PCB	26	0.21	20	0.568
Inductance	97	0.26	4	4.1

Poisson's ratio (v), coefficient of thermal expansion (CTE), and thermal conductivity (K). The initial temperature for both the PCB substrate and the inductor was set at 25 °C.

The initial preset parameters for the transfer molding process of the encapsulation equipment are as follows: filling time of 7 s, melt temperature of 85 °C, mold temperature of 175 °C, filling pressure of 7 MPa, curing time of 120 s, curing pressure of 7 MPa, initial resin conversion rate of 0%, air temperature of 25 °C, and mold opening time of 30 s. Three types of analyses are conducted, including filling analysis, curing analysis, and warpage analysis.

4.4.2   Results and discussion

Figure 4.4 illustrates the variation in filled area over time at 1.5, 3, 5, and 7 s under initial preheating process conditions. It can be observed that the filling speeds of the two gates are nearly identical. The encapsulant flows from both ends of the gates into the runners and subsequently divides into seven channels to fill the encapsulation. The filling rates at the ends of the runners are faster than those in the middle, achieving complete filling of the encapsulation by 7 s. Figure 4.5 reveals that the contribution of the gates from the two fillers is almost equal but not entirely identical. This may be due to the asymmetric placement of inductor components on the substrate, resulting in varying fluid resistance during encapsulation filling.

In transfer molding of encapsulation, resin is prone to occurrences like short shots and flash [23]. Short shots refer to insufficient filling (commonly known as underfilling), whereas flash is its opposite. Neither phenomenon was observed in this study. The resin conversion rate in transfer molding denotes the percentage of resin converted into polymer during the reaction, ensuring the resultant polymer exhibits favorable mechanical properties and chemical stability. It is crucial for the resin conversion rate to be relatively low at the end of the injection stage to avoid issues such as voids and gaps. Figures 4.6 and 4.7 depict the resin conversion rates at the end of filling and after internal curing, respectively. It can be observed that immediately after filling, the resin conversion rate is approximately 1%, indicating

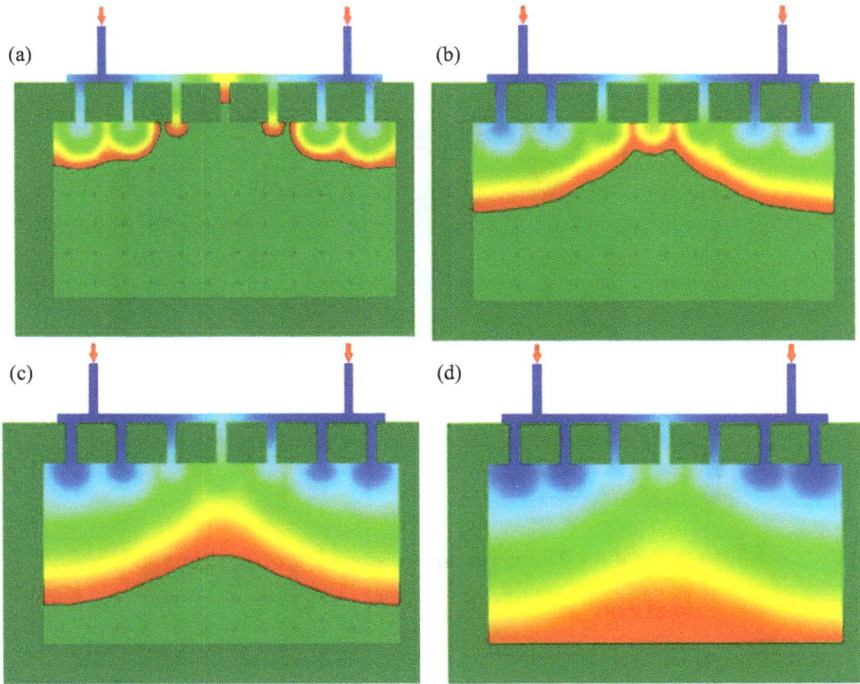

Figure 4.4   Filling area at different times: (a) 1.5 s, (b) 3 s, (c) 5 s, and (d) 7 s

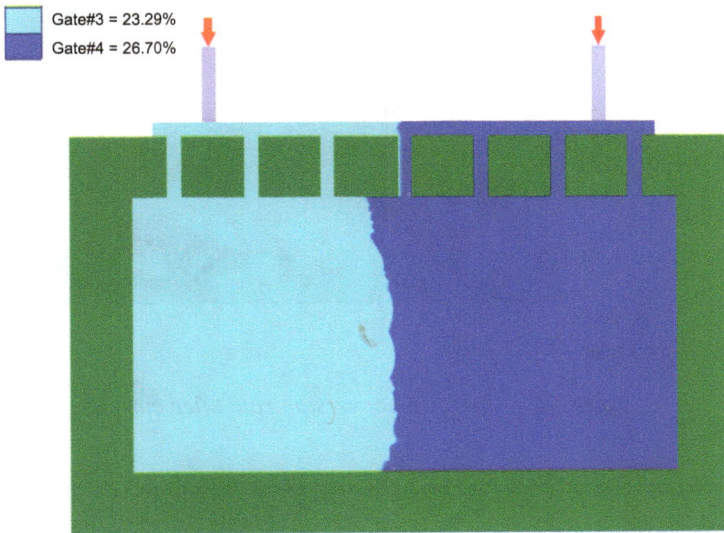

Figure 4.5   Gate contribution of different gates

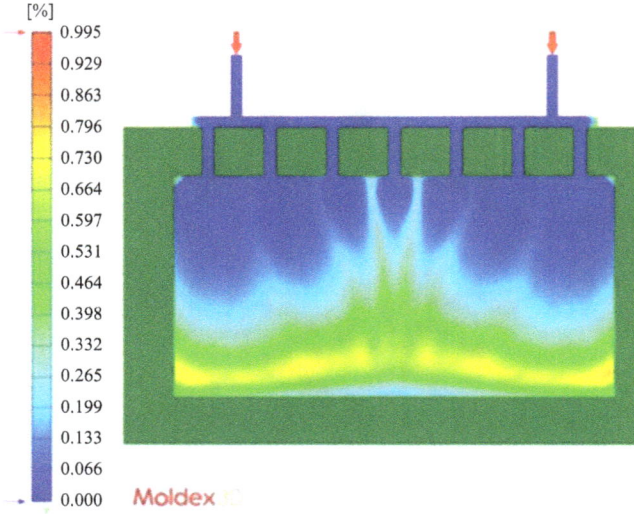

Figure 4.6   The resin conversion rate after filling

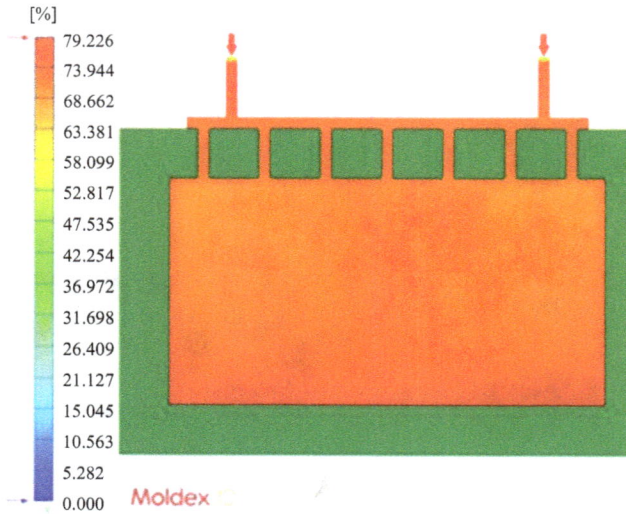

Figure 4.7   The resin conversion rate after curing

satisfactory conversion under the specified process conditions. Following curing, the highest conversion rate reaches 79%, which suggests the need to adjust process parameters such as temperature and curing time to ensure uniform and dense melt distribution.

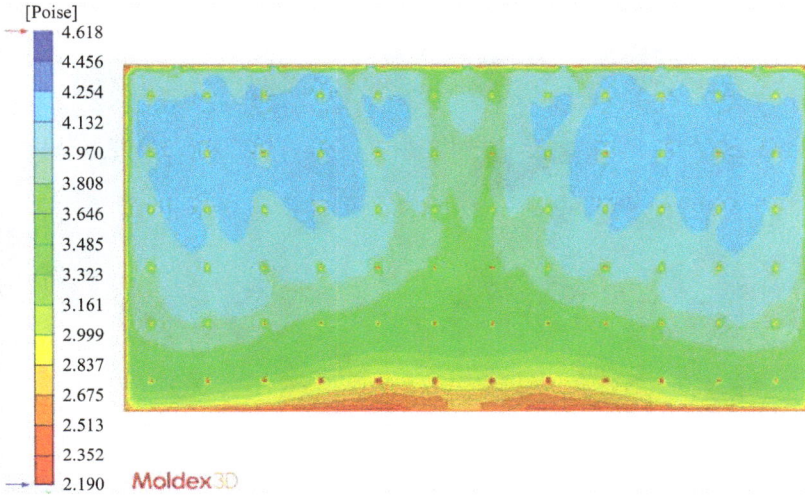

Figure 4.8 Viscosity area distribution after backfilling

In fluid mechanics, viscosity is a measure of the internal resistance to flow within a fluid [24]. It represents the force generated by internal friction between fluid layers, with higher viscosity indicating greater resistance to flow. The value of viscosity depends on fluid properties such as density, temperature, pressure, and chemical composition. Due to the small volume of inductor components, this section presents contour maps extracted from the results, illustrating the distribution of viscosity in the filled region, as shown in Figure 4.8. It can be observed that viscosity is higher near the upper filling port and decreases toward the middle of the bottom section. This is attributed to the relatively lower temperature during the short filling time of the resin, resulting in higher viscosity during filling. As the resin temperature increases, viscosity decreases when filling toward the bottom section.

Shear stress in molten flow is influenced by factors such as viscosity, temperature, and flow velocity, constituting an important parameter in the study of material flow behavior and serving as a source of residual stress within components during the encapsulation process. As depicted in Figure 4.9, shear stress exhibits an inverse relationship with viscosity distribution, decreasing from the upper filling port toward the bottom section where the inductor components are located. This trend is directly attributable to the decreasing viscosity along the flow path.

With the continuous trend toward thinner and smaller electronic packaging, warpage deformation has become an increasingly significant challenge. Excessive warpage can lead to reliability issues such as interfacial delamination and solder joint fractures. As illustrated in Figure 4.10, the vertical board warpage after curing measures 5.637 mm.

In the encapsulation process, the forces experienced during transfer molding vary depending on the position of the product. Simulating inductor stress under

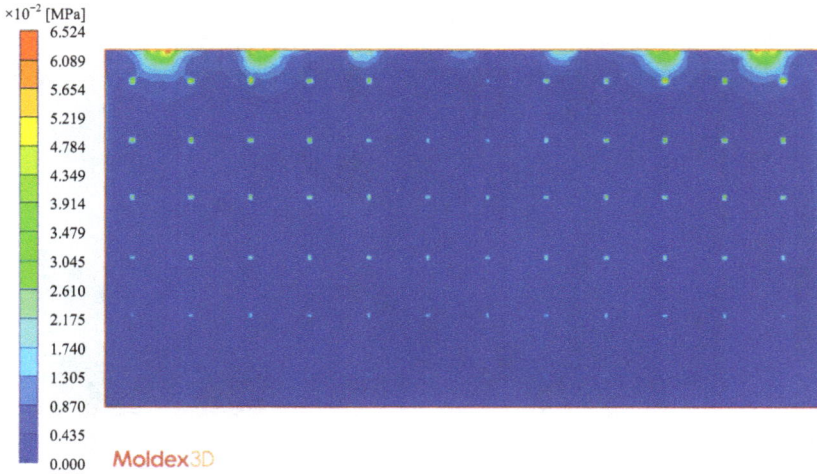

Figure 4.9   Shear stress area distribution after backfilling

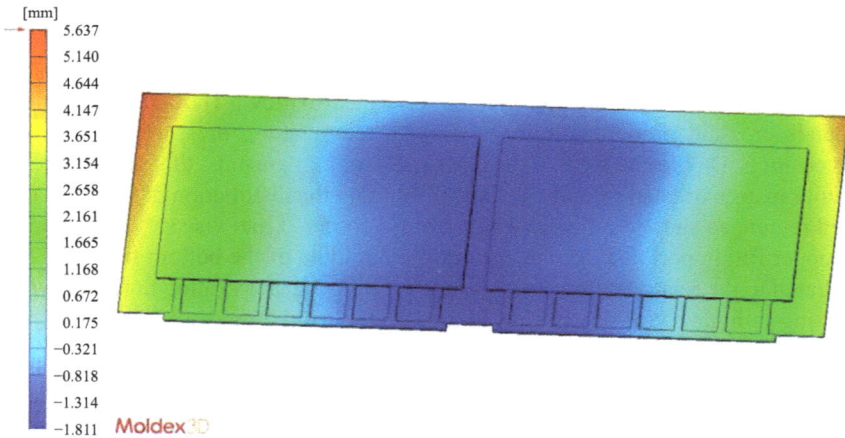

Figure 4.10   The warpage behavior of the substrate after curing

unchanged packaging process parameters, as shown in Figure 4.11, reveals an increasing trend in Mises stress from top to bottom at different locations of the inductor. This phenomenon is primarily influenced by two factors. First, residual stresses from fluids of varying viscosities due to different distances from the filling port affect stress levels. Second, the post-curing thermal expansion coefficient (EMC) induces board warpage, leading to stress generation upon cooling.

In summary, this chapter investigates the influence of transfer molding encapsulation process parameters on mitigating risks such as stress and warpage,

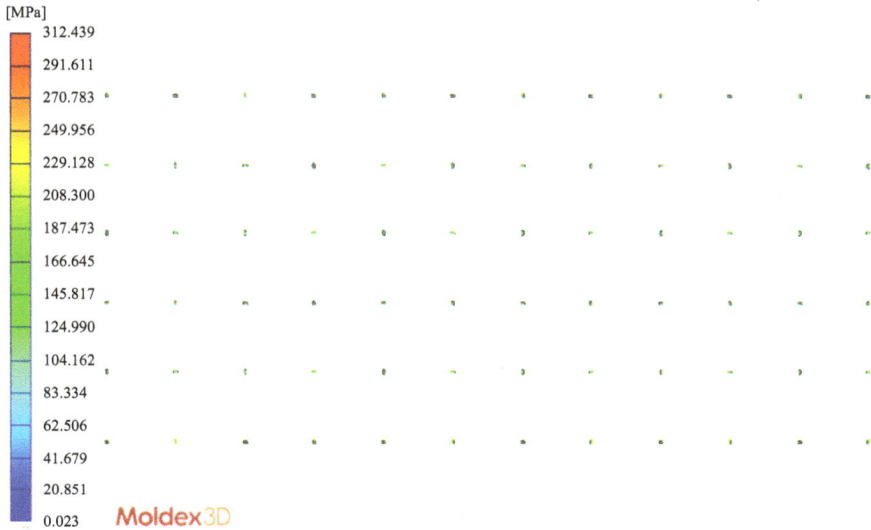

Figure 4.11 The warpage behavior of the substrate after curing

examining various plastic packaging process parameters with warpage and conversion rate as outcome data.

4.4.3 Impact of transfer molding process parameters on warpage

The quality of plastic encapsulation is influenced by various process parameters including filling time, filling pressure, resin melt temperature, curing time, and curing pressure. This section primarily investigates and discusses the effects of various process parameters on the transfer molding encapsulation process. Filling time is a critical parameter in transfer molding processes, typically controlled by adjusting the descent rate of the plunger in the molding machine setup. Insufficient filling time can result in incomplete filling, with resin entering the mold cavity at a higher flow rate, thereby subjecting the components to greater impact and shear forces and potentially causing short shots and air bubbles. Conversely, excessive filling time increases production costs, and lower resin temperatures with higher viscosity can lead to defects such as short shots and deformation.

Figure 4.12 illustrates the impact of filling time on transfer molding under default conditions of other process parameters. It can be observed that when the filling time is 5 s or more, there is a trend of increasing conversion rates post-filling and post-curing, as well as increasing vertical warpage and stress. However, when the filling time is too short, coupled with an insufficient filling pressure of 7 MPa and a small filling port, incomplete filling and short shots may occur.

Figure 4.13 shows that with an actual filling time of 3.896 s and a maximum filling pressure of 4.9 MPa, the filling was incomplete. Therefore, for plastic

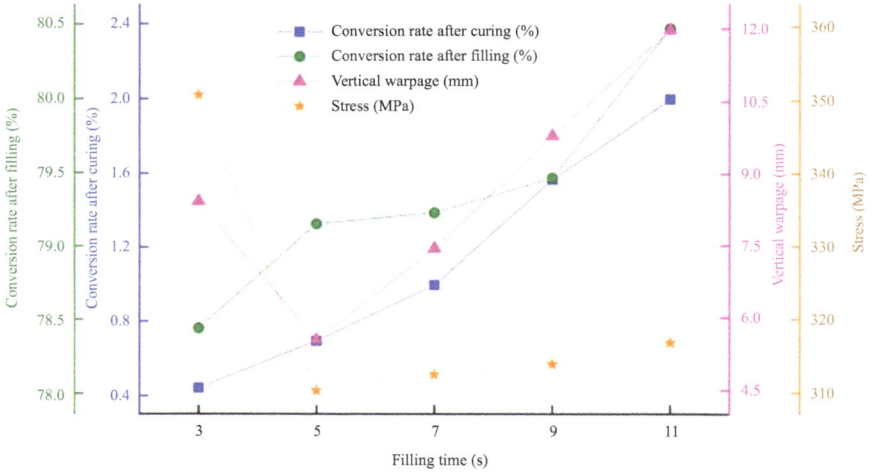

Figure 4.12   The influence of filling time on transfer molding

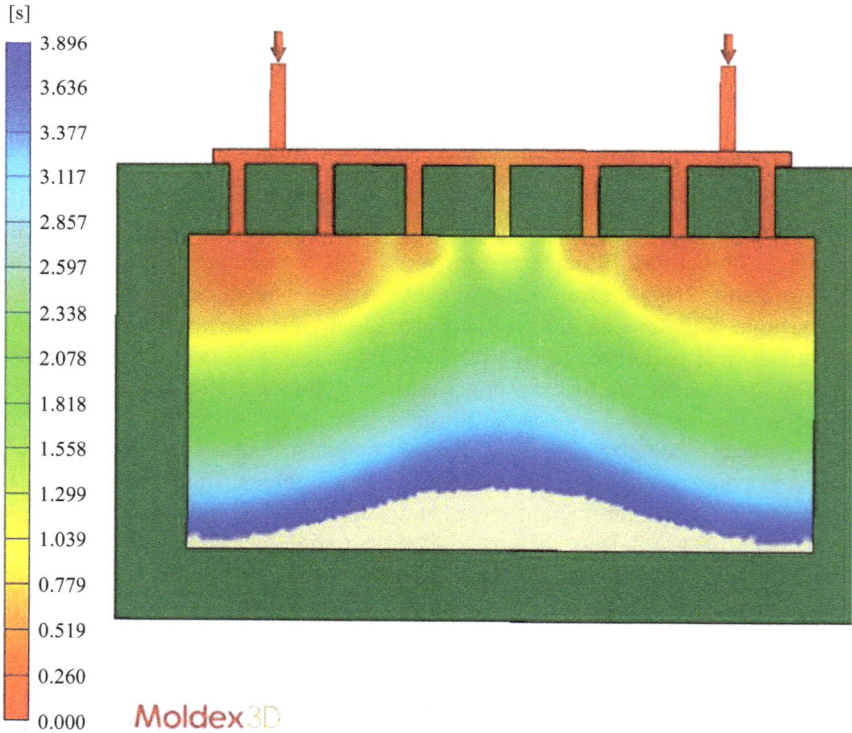

Figure 4.13   The filling time is short and incomplete

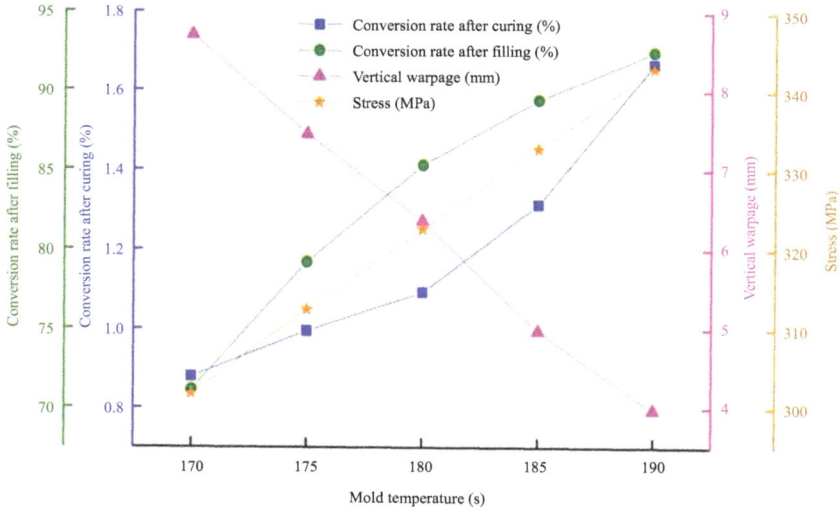

Figure 4.14   The influence of mold temperature on transfer molding

encapsulation process parameters, it is necessary to ensure an adequate filling time to achieve complete filling of the encapsulation body.

Mold temperature influences molding time and quality: Higher mold temperatures shorten resin solidification time, accelerate molding speed, and cause rapid solidification at the mold contact surface, leading to surface defects. However, excessively high temperatures may result in rapid resin curing without complete mold filling, thereby compromising the quality of plastic encapsulation. Conversely, too low mold temperatures may cause molded parts to adhere to the mold, making demolding difficult. The impact of mold temperature on transfer molding is illustrated in Figure 4.14. It is evident that mold temperature significantly affects resin conversion rate; a difference of nearly 20% in conversion rate is observed between curing temperatures of 170 °C and 190 °C. Moreover, warpage increases markedly with higher mold temperatures, while stress decreases by approximately 41 MPa.

Melt temperature affects the flowability, viscosity, and molding time of molten resin. A higher melt temperature results in lower viscosity, better flowability of the molten resin, and shorter solidification time. However, resin at high temperatures is more prone to defects such as shrinkage voids or bubbles. Conversely, excessively low temperatures lead to inadequate flowability, resulting in defective or incomplete molded parts.

Resin temperature's impact on transfer molding is depicted in Figure 4.15. It is evident from the figure that with increasing resin melt temperature, the conversion rate, warpage, and stress after filling and curing all increase. Moreover, under high melt temperature conditions, there is a significant increase in resin conversion rate post-curing.

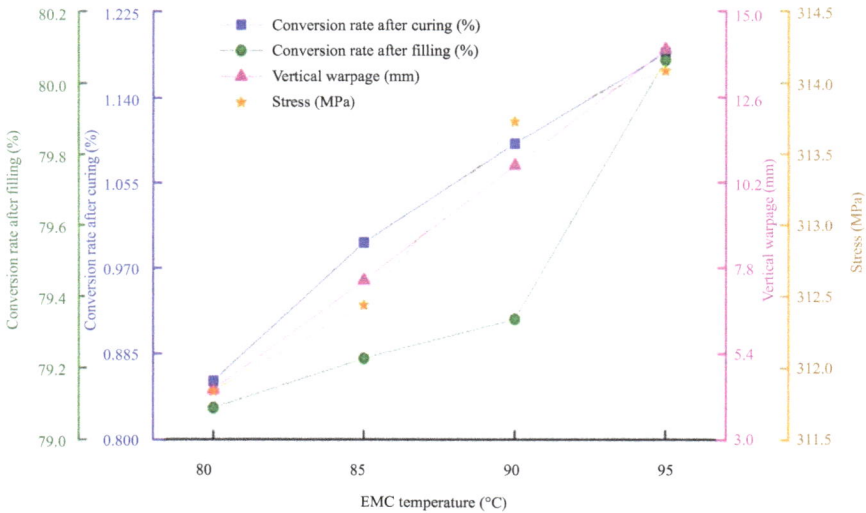

Figure 4.15   The influence of EMC temperature on transfer molding

Curing time is a critical parameter in the transfer molding process; the longer the curing time, the longer the molding cycle. Therefore, selecting an appropriate curing time can enhance production efficiency and capacity. Additionally, it affects the surface quality of the encapsulation. Excessive curing time may result in wrinkles or molten deformation on the encapsulation surface. Conversely, too short a curing time may lead to uneven surfaces on the encapsulation. Altering only the curing time process parameter has a significant impact on the post-curing conversion rate, showing a rising trend followed by stabilization as time increases. However, it has minimal effect on post-filling conversion rate, warpage, and stress, as illustrated in Figure 4.16.

4.4.4   Taguchi orthogonal experimental design and analysis

In practical encapsulation processes, multiple process parameters typically interact and collectively influence outcomes. To investigate the effects of process parameters on encapsulation conversion rate, warpage, and stress, this study employed statistical analysis software MINITAB and utilized the Taguchi orthogonal experimental design method for experimental design and analysis [25]. In the analysis and design research of electronic packaging reliability, Taguchi orthogonal experiments have been widely adopted. In this subsection, an L9 (3^4) standard orthogonal array was established using the Taguchi orthogonal experimental design method, conducting a total of nine simulation experiments. Design variables significantly impacting encapsulation results across multiple parameters were identified, and optimal combinations of various design variables were determined.

Given the need to assess multiple encapsulation outcomes for process parameter evaluation, a comprehensive weighted scoring was applied to normalize

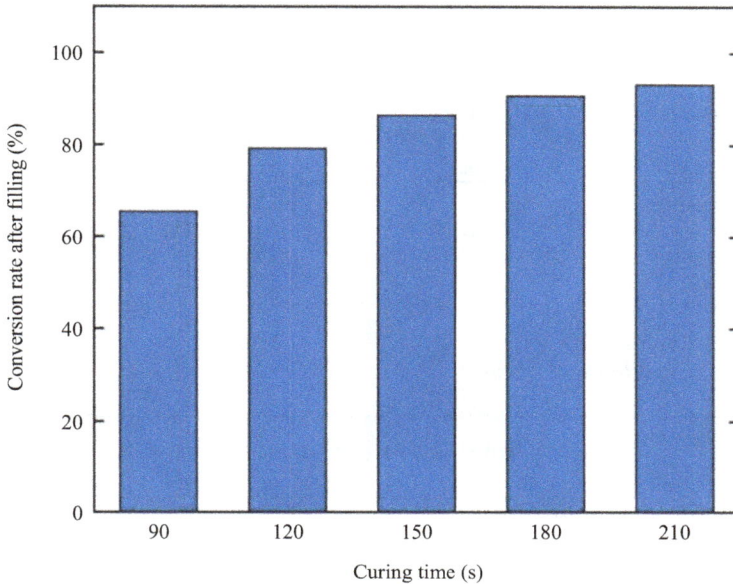

Figure 4.16   The influence of cure time on transfer molding

simulation results, calculating the optimal solution for the encapsulation process, as illustrated in (4.24) and (4.25). Considering the critical impact of warpage behavior on encapsulation processes and the necessity to address product stress, priority was given to these two simulation outcomes for their favorable effects on encapsulation processes in this simulation experiment. To facilitate comparative analysis and synthesis of weighted results, an equal weight ratio of 0.5 was assigned to warpage and stress values

$$Y_K = 100 \times \frac{K_{\text{Max}} - K_i}{K_{\text{Max}} - K_{\text{Min}}} \tag{4.24}$$

$$Y = \sum_{i=1}^{i=n} W_i K_i \tag{4.25}$$

The experimental design analysis selected four encapsulation process parameters—filling time, resin melt temperature, mold temperature, and curing time—as design variables, while material parameters and structural parameters remained fixed at their initial design values. The design variables and their respective levels are shown in Table 4.3, where each design variable has three levels.

According to the standard orthogonal array analysis, the design combinations of various transfer molding process parameters and their simulation results are presented in Table 4.4. It can be observed that the first experimental factor combination achieves the highest comprehensive weighted score. However, further

Table 4.3   Design variables and levels of warpage and conversion analysis for package

Design variables	Factors	Level		
		1	2	3
EMC temperature (°C)	A	80	85	90
Filling time (s)	B	5	7	9
Mold temperature (°C)	C	175	180	185
Cure time (s)	D	120	150	180

Table 4.4   Experimental results

Factor and level				Conversion (%)	Warpage (mm)	Stress (MPa)	Weighted score
A	B	C	D				
1 80	5	175	120	78.989	3.774	311.530	96.757
2 80	7	180	150	90.519	3.222	322.083	77.730
3 80	9	185	180	95.404	4.748	335.223	41.034
4 85	5	180	180	93.403	3.500	319.432	81.691
5 85	7	185	120	89.449	4.974	332.670	45.094
6 85	9	175	150	86.710	9.799	313.829	56.506
7 90	5	185	150	93.243	5.178	330.106	49.306
8 90	7	175	180	90.909	10.684	313.730	51.515
9 90	9	180	120	85.380	11.732	324.641	22.331

Table 4.5   Range analysis of conversion rate after curing of design variables

		A	B	C	D
Conversion rate after curing	Level 1	88.304	88.545	85.543	84.606
	Level 2	89.854	90.299	89.767	90.157
	Level 3	89.851	89.165	92.699	93.245
	Delta	4.550	1.754	7.156	8.639
	Rank	4	3	2	1
	Opt. level	A2	B2	C3	D3

analysis is required to determine whether combination 1 represents the optimal process solution.

The range analysis of the above-mentioned experimental results is presented in Table 4.5. In this table, $K1$, $K2$, and $K3$ reflect the degree to which the levels of each factor influence the experimental outcomes. In the weighted scoring criterion, a higher K value indicates better overall performance, whereas in the remaining two experimental results, the relationship is reversed. Based on the range R values, the impact of each factor on the conversion rate after curing is determined as follows: $D > C > B > A$. The curing time exhibits the most significant influence on the conversion rate after curing. The optimized structural combination is [A2, B2, C3, D3], corresponding to a resin melt temperature of 85 °C, filling time of 7 s, mold temperature of 185 °C, and curing time of 180 s, which aligns with the theoretical findings.

Given that the results presented in the previous subsection have already indicated that the curing time has almost no effect on warpage and stress, the analysis of the results disregards the data from the curing time analysis and directly selects option D3 with the largest conversion rate after curing. Table 4.6 presents the range analysis of the design variable warpage. Based on the range R value, the extent of influence of each factor on warpage is determined as follows: $A > B > C$. The most significant effect on warpage is observed for the resin melt temperature, with the optimized structural combination being [A1, B1, C3, D3], corresponding to a resin melt temperature of 80 °C, filling time of 5 s, mold temperature of 185 °C, and curing time of 180 s.

Table 4.7 presents the range analysis of the design variable stress. Based on the range R value, the degree of influence of each factor on stress is determined as follows: $C > B > A$. The mold temperature exhibits the most significant impact on stress, with the optimized structural combination being [A2, B1, C1, D3], corresponding to a resin melt temperature of 85 °C, filling time of 5 s, mold temperature of 175 °C, and curing time of 180 s.

Table 4.6   Range analysis of design variable warpage value

		A	B	C	D
Warpage (mm)	Level 1	3.915	4.151	8.086	6.827
	Level 2	6.091	6.293	6.151	6.066
	Level 3	9.198	8.760	4.967	6.311
	Delta	5.283	4.609	3.119	0.760
	Rank	1	2	3	4
	Opt. level	A1	B1	C3	D3

Table 4.7   Range analysis of design variable stress

		A	B	C	D
Stress (MPa)	Level 1	322.9	320.4	313.0	322.9
	Level 2	322.0	322.8	322.1	322.0
	Level 3	322.8	327.6	332.7	322.8
	Delta	0.968	4.208	1.9637	0.941
	Rank	3	2	1	4
	Opt. level	A2	B1	C1	D3

Table 4.8   Range analysis of comprehensive weighted scores of design variables

		A	B	C	D
Weighted composite scores	Level 1	71.840	75.918	68.259	54.727
	Level 2	61.097	58.113	60.584	61.181
	Level 3	41.051	39.957	45.145	58.080
	Delta	30.790	35.961	23.115	6.453
	Rank	2	1	3	4
	Opt. level	A1	B1	C1	D3

The above two sets of experiments show completely opposite primary influencing factors on warpage and stress. Range analysis of the weighted composite scores is presented in Table 4.8, indicating that filling time has the most significant impact on the results when considering all factors together, followed by resin temperature and mold temperature. The optimal transfer molding process parameters are [A1, B1, C1, D3], corresponding to a resin melt temperature of 80 °C, filling time of 5 s, mold temperature of 175 °C, and curing time of 180 s. Under these secondary molding conditions, the conversion rate after curing is 90.818%, warpage is 3.774 mm, and stress is 311.530 MPa. These can serve as preset parameters for the transfer molding machine. However, in practical transfer molding processes, further parameter adjustments based on actual conditions are necessary to ensure the reliability and safety of the sealed products.

4.5   Chapter summary

In this chapter, a comprehensive study of the transfer molding process was conducted using advanced three-dimensional mold flow analysis software. This software simulation provides a valuable tool for understanding and predicting the behavior of the resin during the molding process, offering theoretical guidance that helps optimize the actual encapsulation process before any physical prototypes are created. The primary goal was to explore the impact of various encapsulation process parameters on key performance metrics such as resin conversion rate, substrate warpage, and stress distribution within the product. These parameters are critical in determining the quality and reliability of the final encapsulated product, making it essential to understand their individual and combined effects.

A systematic approach was adopted to study these parameters in detail. Initially, a single-variable method was employed to isolate and investigate the effects of each individual process parameter on the outcomes. The key parameters examined included resin melt temperature, filling time, filling pressure, curing time, and mold temperature. Each of these factors was varied independently to observe how they influenced the resin flow, conversion rate, and mechanical properties of the encapsulated product. It was found that each of these parameters had a varying degree of influence on the final results, with some having a more pronounced impact on resin flow and others playing a significant role in controlling the curing and stress distribution. For instance, the resin melt temperature had a direct influence on the flowability of the resin, while mold temperature and curing time were critical in determining the degree of polymerization and the resulting stresses in the encapsulated product.

Furthermore, the combined effects of these parameters were explored, revealing that their interactions could significantly alter the encapsulation results. For example, an increase in mold temperature in combination with a longer curing time could result in reduced warpage but might also lead to higher residual stresses. These findings underscored the complexity of the transfer molding process, where multiple parameters must be balanced to achieve the desired product characteristics.

To determine the optimal set of process parameters that would minimize warpage and stress while maximizing the resin conversion rate, a more advanced approach was adopted. The Taguchi orthogonal method, a statistical design of experiments technique, was employed to analyze the effects of multiple parameters simultaneously. This method is particularly useful in situations where there are numerous factors to consider, as it allows for the evaluation of several factors with a limited number of experiments. By applying the Taguchi method, the interaction effects between parameters could be more accurately assessed, leading to a more efficient identification of optimal process conditions.

After conducting the experiments and analyzing the results, the optimal transfer molding process parameters were determined. The chosen set of parameters was found to strike an effective balance between minimizing warpage and stress while ensuring an efficient molding process. The most suitable predefined parameters were identified as follows: a resin melt temperature of 85 °C, a filling time of 5 s, a mold temperature of 175 °C, and a curing time of 180 s. These values were selected based on a set of weighted objectives, where warpage and stress were each given equal importance (0.5). This approach ensured that the final product would meet the necessary structural integrity requirements without excessive deformation or stress buildup, leading to higher reliability and performance of the encapsulated product.

The results of this study provide valuable insights into the optimal conditions for the transfer molding process, demonstrating the effectiveness of using simulation tools and statistical methods like the Taguchi method for process optimization. By accurately predicting the influence of process parameters on the resin flow, warpage, and stress, manufacturers can improve product quality, reduce the need for trial-and-error physical testing, and streamline the design and production processes. Ultimately, this work offers a more efficient and reliable method for optimizing transfer molding encapsulation, which can be applied to a wide range of industries, from electronics to automotive manufacturing.

References

[1]   S.K. Lo, T.Y. Jian, Z.Y. Li, *et al.*, Optimization of process parameters for molding simulation of open cavity QFN, in: *2022 IEEE 24th Electronics*

Packaging Technology Conference (EPTC), IEEE, Singapore, Singapore, 2022: pp. 655–659.

[2]   A.T. Kian Siang, Transfer molding wire sweep improvement in QFP packages by mold plunger tips design, in: *2022 IEEE 39th International Electronics Manufacturing Technology Conference (IEMT)*, IEEE, Kuala Lumpur, Putrajaya, Malaysia, 2022: pp. 1–4.

[3]   Y. Wen, C. Chen, Y. Ye, *et al.*, Advances on thermally conductive epoxy-based composites as electronic packaging underfill materials: A review, *Advanced Materials* 34 (2022) 2201023.

[4]   F.C. Ng and M.A. Abas, Underfill flow in flip-chip encapsulation process: A review, *Journal of Electronic Packaging* 144 (2022) 010803.

[5]   P.J. Schubel, N.A. Warrior, and C.D. Rudd, Surface quality prediction of thermoset composite structures using geometric simulation tools, *Plastics, Rubber and Composites* 36 (2007) 428–437.

[6]   M. Haider, P. Hubert, and L. Lessard, Cure shrinkage characterization and modeling of a polyester resin containing low profile additives, *Composites Part A: Applied Science and Manufacturing* 38 (2007) 994–1009.

[7]   J. Tamil, S.H. Ore, K.Y. Gan, *et al.*, Molding flow modeling and experimental study on void control for flip chip package panel molding with molded underfill technology, *Journal of Microelectronics and Electronic Packaging* 9 (2012) 19–30.

[8]   J. Huang, X. Huang, K. Shi, X. Cheng, and Y. Jiang, Mold flow analysis and optimization in injection molding process for semiconductor packages, in: *2022 23rd International Conference on Electronic Packaging Technology (ICEPT)*, IEEE, Dalian, China, 2022: pp. 1–5.

[9]   Y.K. Kim, I.S. Park, and J. Choi, Warpage mechanism analyses of strip panel type PBGA chip packaging, *Microelectronics Reliability* 50 (2010) 398–406.

[10]   J. De Vreugd, K.M.B. Jansen, A. Xiao, *et al.*, Advanced viscoelastic material model for predicting warpage of a QFN panel, in: *2008 58th Electronic Components and Technology Conference*, IEEE, Lake Buena Vista, FL, USA, 2008: pp. 1635–1640.

[11]   J. Luan, Integrated methodology for warpage prediction of IC packages, in: *2006 Thirty-First IEEE/CPMT International Electronics Manufacturing Technology Symposium*, IEEE, Petaling Jaya, Malaysia, 2006: pp. 143–149.

[12]   K. Cho and I. Jeon, Numerical analysis of the warpage problem in TSOP, *Microelectronics Reliability* 44 (2004) 621–626.

[13]   T.-C. Chiu, C.-L. Gung, H.-W. Huang, and Y.-S. Lai, Effects of curing and chemical aging on warpage—Characterization and simulation, *Transactions on Device and Materials Reliability* 11 (2011) 339–348.

[14]   T.-C. Chiu, H.-W. Huang, and Y.-S. Lai, Warpage evolution of overmolded ball grid array package during post-mold curing thermal process, *Microelectronics Reliability* 51 (2011) 2263–2273.

[15]   D. Adolf and R. Chambers, Verification of the capability for quantitative stress prediction during epoxy cure, *Polymer* 38 (1997) 5481–5490.

[16] D. Adolf and J.E. Martin, Calculation of stresses in crosslinking polymers, *Journal of Composite Materials* 30 (1996) 13–34.

[17] D.G. Yang, K.M.B. Jansen, L.G. Wang, *et al.*, Micromechanical modeling of stress evolution induced during cure in a particle-filled electronic packaging polymer, *IEEE Transactions on Components, Packaging and Manufacturing Technology Information* 27 (2004) 676–683.

[18] D.B. Adolf, J.E. Martin, R.S. Chambers, S.N. Burchett, and T.R. Guess, Stresses during thermoset cure, *Journal of Materials Research* 13 (1998) 530–550.

[19] S.L. Simon, G.B. Mckenna, and O. Sindt, Modeling the evolution of the dynamic mechanical properties of a commercial epoxy during cure after gelation, *Journal of Applied Polymer Science* 76 (2000) 495–508.

[20] T.A. Bogetti and J.W. Gillespie, Process-induced stress and deformation in thick-section thermoset composite laminates, *Journal of Composite Materials* 26 (1992) 626–660.

[21] C. Koplin, R. Jaeger, and P. Hahn, A material model for internal stress of dental composites caused by the curing process, *Dental Materials* 25 (2009) 331–338.

[22] Z. Zhang and C.P. Wong, Recent advances in flip-chip underfill: Materials, process, and reliability, *IEEE Transactions on Advanced Packaging* 27 (2004) 515–524.

[23] C.Y. Shen, X.R. Yu, L.X. Wang, and Z. Tian, Gate location optimization of plastic injection molding, *CIESC Journal* 55 (2004) 445.

[24] R.W. Fox, A.T. McDonald, and J.W. Mitchell, *Fluid Mechanics*, Asia Edition, 10th Edition, Wiley.

[25] I. Syahzaqi, H.B. Rochmanto, and M. Ahsan, Selecting optimal process parameters of Al_2O_3/C composite using GRA with PCA and Taguchi's QLF approach, *Barekeng: Journal of Mathematics and Its Applications* 16 (2022) 1039–1050.

Chapter 5

Structural integration simulation under vibration for PBGA packaging

5.1 Introduction

As electronic devices continue to advance toward miniaturization, thinness, and multifunctionality, the operational environments in which these devices are deployed have become increasingly demanding. These harsher environments exacerbate the reliability challenges associated with plastic ball grid array (PBGA) packaging, which has become a widely used technology in modern electronic devices. In PBGA packaging, the solder joints that connect the package to the printed circuit board (PCB) are critical for both mechanical support and electrical interconnection. However, these solder joints are also the most vulnerable components, making their reliability a crucial factor in ensuring the overall performance and longevity of the device. The failure of a single solder joint can lead to the failure of the entire component, or, in some cases, the failure of the entire system. This highlights the importance of addressing solder joint reliability when designing PBGA packages, especially in devices subjected to extreme operating conditions.

PBGA packaging is commonly used in devices that encounter a variety of service environments, including temperature cycling, vibration, mechanical shock, and high-temperature, high-humidity conditions. According to the US Air Force statistics, about 55% of all electronic device failures are attributed to temperature-related loads, while approximately 20% are due to vibration and shock, with a significant portion of these failures resulting from solder joint issues [1]. Given these statistics, it is clear that understanding and improving the reliability of solder joints under such stressors is essential for enhancing the robustness of PBGA-packaged devices. Research into the mechanical performance and reliability of PBGA packaging, particularly under vibrational loads, is therefore of great importance. By identifying the failure mechanisms of solder joints and addressing potential defects, researchers can provide valuable insights that guide the development of more reliable PBGA packaging solutions.

One of the key challenges in PBGA packaging is the inevitable exposure to vibrational loads during the operational life of the device. Vibrations, in particular, play a significant role in the degradation of solder joint integrity over time. The effects of vibration are complex and can be broadly categorized into two main types: harmonic vibrations and random vibrations. Harmonic vibrations, which occur at specific frequencies, are relatively straightforward to analyze and predict. However, in real-world scenarios, devices are more commonly subjected to random vibrations, which vary in intensity, frequency, and direction. These random vibrations are more difficult to model and predict due to their stochastic nature, yet they are the most representative of the operating conditions that PBGA-packaged devices typically face.

In this chapter, numerical simulations are used to investigate the effects of both harmonic sweep and random vibrations on the stress distribution in PBGA package solder joints. By varying the chip's position on the PCB, the study explores how these vibrations influence the mechanical stresses experienced by the solder joints. The simulations allow for the identification of critical solder joint locations, which are most susceptible to failure, based on the stress distribution. This information is crucial for improving the design of PBGA packages, enabling engineers to focus on reinforcing vulnerable areas and potentially extending the life of the package. Furthermore, by comparing the effects of harmonic sweep vibration and random vibration, the study sheds light on the differences in stress profiles generated by these two types of vibrations, thereby offering insights into the relative risks posed by each type.

To enhance the reliability of PBGA packaging under vibrational loads, the chapter also investigates various factors that significantly affect solder joint stress. Through experimental design and sensitivity analysis, optimal parameter combinations are determined, which can be employed to reduce the stress on solder joints. These parameters include factors such as PCB layout, solder material properties, ball size, and package geometry. By optimizing these design parameters, it is possible to reduce the likelihood of solder joint failure and improve the overall performance and lifespan of PBGA-packaged devices in harsh operating conditions.

This chapter highlights the critical role of solder joint reliability in the performance of PBGA packaging, especially under vibrational loads. By employing numerical simulations and experimental analysis, the study provides valuable insights into the stress distribution and failure mechanisms of solder joints, enabling the development of more robust PBGA packaging solutions. The findings have the potential to guide the future design and manufacturing of electronic devices, ensuring they can withstand the increasingly harsh conditions of modern operating environments.

5.2 Mechanical vibration theory

5.2.1 *Theory of simple harmonic vibration*

For the single-degree-of-freedom system shown in Figure 5.1, under the action of harmonic excitation, the differential equation of motion is [2]

$$m\ddot{x} + c\dot{x} + kc = F \sin \omega t \tag{5.1}$$

where m is the mass, c is the damping, and k is the stiffness coefficient.

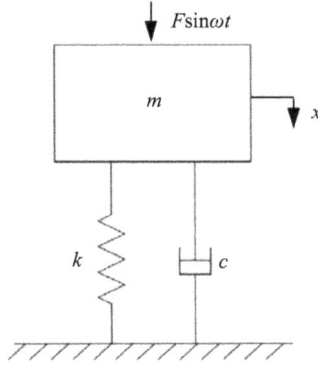

Figure 5.1   Single-degree-of-freedom system under harmonic excitation

To obtain the motion characteristics of the system, it is necessary to determine the solution to (5.1). The general solution to the homogeneous equation of (5.1) is

$$x_1 = e^{-\xi\omega_n t}(A \cos \omega_d t + B \sin \omega_d t) \tag{5.2}$$

where ξ is the damping ratio, $\xi = c/2\sqrt{mk}$, ω_n is the natural frequency, $\omega_n = \sqrt{k/m}$, ω_d is the damped resonant frequency, $\omega_d = \sqrt{1 - 2\xi^2}\omega_n$, and A and B are constants.

The particular solution to the non-homogeneous equation (5.1) is

$$x_2 = \frac{F \sin(\omega t - \varphi)}{\sqrt{(k - m\omega^2)^2 + (c\omega)^2}} \tag{5.3}$$

where φ is the phase angle and $\varphi = \arctan[c\omega/(k - m\omega^2)]$.

Thus, the response of the single-degree-of-freedom system to harmonic excitation is

$$x = x_1 + x_2 = e^{-\xi\omega_n t}(A \cos \omega_d t + B \sin \omega_d t) + \frac{F \sin(\omega t - \varphi)}{\sqrt{(k - m\omega^2)^2 + (c\omega)^2}} \tag{5.4}$$

5.2.2   Random vibration theory

For the single-degree-of-freedom system shown in Figure 5.2, under random excitation, the system's response is computed using convolution [3]

$$x(t) = f(t) * h(t) = \int_{-\infty}^{\infty} f(\tau)h(t - \tau)d\tau \tag{5.5}$$

where $h(t)$ is the system's unit impulse response.

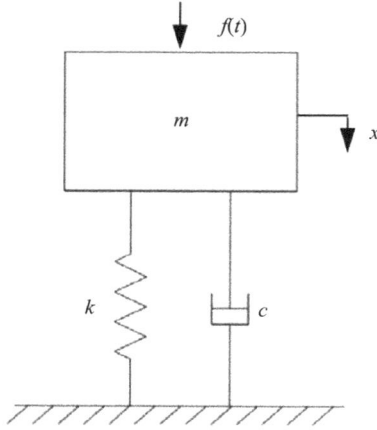

Figure 5.2   Single degree of freedom system under random excitation

Taking the Fourier transform of (5.5)

$$X(\omega) = H(\omega)F(\omega) \tag{5.6}$$

Taking the complex conjugate of both sides of (5.6) yields

$$X^*(\omega) = H^*(\omega)F^*(\omega) \tag{5.7}$$

Combining with the power spectral density function $S(\omega)$ in (5.8), multiply both sides of (5.6) and (5.7), respectively, to obtain the relationship between the random excitation and response of a single-input-single-output system

$$S(\omega) = \frac{1}{T}|X(\omega)|^2 \tag{5.8}$$

$$S_x(\omega) = |H(\omega)|^2 S_f(\omega) \tag{5.9}$$

The statistical regularity of random excitation $f(t)$ is represented by power spectral density $S_f(\omega)$, and the mean square response can be calculated using the following equation:

$$\bar{x} = E[x^2(t)] = \frac{1}{2\pi}\int_{-\infty}^{\infty}|H(\omega)|^2 S_f(\omega)d\omega \tag{5.10}$$

The mean square value of the response reflects the average energy level of the vibration and can be used as a basis for calculating the fatigue strength of machine structures. In the study of random vibration responses, the power spectral density is used to represent the statistical characteristics of the random vibration process across different frequencies in the frequency domain. Due to the convenience of the relationship between the input and output power spectra of the system, frequency domain analysis is commonly used for random vibrations. Therefore, a brief introduction to power spectral density follows [4].

Assuming $x(t)$ be a stationary random process that is a non-periodic real function of time and satisfies the following conditions: (1) $\int_{-\infty}^{+\infty} x(t)dt < +\infty$, $x(t)$ is absolutely integrable; (2) $x(t)$ has a finite number of first-kind discontinuities and a finite number of extrema within any interval, satisfying the Dirichlet condition. Then, the Fourier transform of $x(t)$ exists and is given by

$$F_x(\omega) = \int_{-\infty}^{\infty} x(t)e^{-i\omega t}dt \tag{5.11}$$

It is called the spectral density, which reflects the distribution of various frequency components in $x(t)$.

An arbitrary sample function of a random process generally does not satisfy the condition of absolute integrability required for the Fourier transform and thus cannot be directly transformed. However, after imposing certain restrictions on its sample function, the Fourier transform can exist. The simplest approach is to use a truncation function and take any segment of length $2T$ within $x(t)$

$$x(t) = \begin{cases} x(t), |t| \leq T \\ 0, |t| > T \end{cases} \tag{5.12}$$

The function $X_T(t)$ is called the truncation of $x(t)$, when T is finite, the truncation on function $X_T(t)$ satisfies the condition absolute integrability, and its Fourier transform exists

$$F_x(\omega, T) = \int_{-\infty}^{\infty} X_T(t)e^{-i\omega t}dt \tag{5.13}$$

The corresponding Parseval's theorem is

$$\int_{-T}^{T} X_T^2(t)dt = \frac{1}{2\pi} \int_{-\infty}^{\infty} |F_x(\omega, T)|^2 d\omega \tag{5.14}$$

Dividing both sides by $2T$ yields

$$\frac{1}{2T} \int_{-T}^{T} X_T^2(t)dt = \frac{1}{4\pi T} |F_x(\omega, T)|^2 d\omega \tag{5.15}$$

Let $T \to +\infty$, $x(t)$ can be expressed as the following in $(-\infty, +\infty)$

$$\lim_{T \to \infty} \frac{1}{2T} \int_{-T}^{T} X_T^2(t)dt = \frac{1}{2\pi} \int_{-\infty}^{\infty} \frac{1}{2T} |F_x(\omega, T)|^2 d\omega \tag{5.16}$$

The integrand on the right side of (5.17) is called the average power spectral density of the function $x(t)$ and is denoted as

$$S_x(\omega) = \lim_{T \to \infty} \frac{1}{2T} |F_x(\omega, T)|^2 \tag{5.17}$$

For a general stationary random process $X(t)$, $-\infty < t < +\infty$, from (5.13) and (5.15), it follows that

$$F_x(\omega, T) = \int_{-T}^{T} X(t)e^{i\omega t}dt \tag{5.18}$$

$$\frac{1}{2T}\int_{-T}^{T} X^2(T)dt = \frac{1}{4\pi T}\int_{-\infty}^{\infty} |F_x(\omega, T)|^2 d\omega \tag{5.19}$$

Taking the limit of the mean on the left side of (5.19), the average power of the stationary random process $X(t)$ is obtained as

$$\lim_{T\to\infty} E\left[\frac{1}{2T}\int_{-T}^{T} X^2(t)dt\right] \tag{5.20}$$

For a stationary random process, where the mean square value is constant ψ^2, swapping the order of integration and mean operation in (5.20) yields

$$\lim_{T\to\infty} E\left[\frac{1}{2T}\int_{-T}^{T} X^2(t)dt\right] = \lim_{T\to\infty}\frac{1}{2T}\int_{-T}^{T} E[X^2(t)]dt = \psi_X^2 \tag{5.21}$$

From (5.19) and (5.21), it can be concluded that

$$\psi_X^2 = \frac{1}{2\pi}\int_{-\infty}^{+\infty} \lim_{T\to\infty}\frac{1}{2\pi}E[|F_X(\omega, T)^2|]d\omega \tag{5.22}$$

The integrand is called the power spectral density of the stationary random process $X(t)$, denoted as $S_x(\omega)$, and we have

$$S_x(\omega) = \lim_{T\to\infty}\frac{1}{2T}E[|F_x(\omega, T)|^2] \tag{5.23}$$

5.3 Establishment of numerical model for PBGA packaging

Due to the small size of PBGA package solder joints and their mounting on the PCB, conventional stress measurements pose challenges. Numerical simulation offers a potential solution to this problem. To obtain an effective finite-element model, modal parameters, acceleration, and displacement responses of the PBGA package finite-element model under harmonic sweep vibration loading are first obtained through finite-element analysis. These results are then compared with experimental data to refine the numerical model, ultimately resulting in a feasible numerical model.

5.3.1 Geometric characteristics of PBGA packaging

The PBGA packaging sample studied in this chapter is shown in Figure 5.3. The sample contains 12 PBGA-packaged chips and is primarily composed of six parts:

Figure 5.3 Layout of the test vehicle

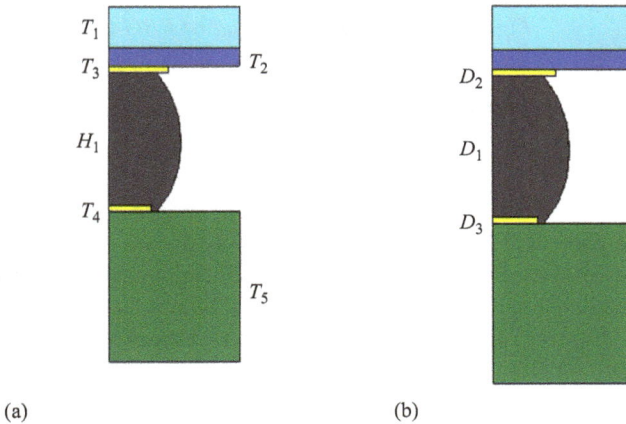

(a)   (b)

Figure 5.4 Geometry dimension of solder joint cell: (a) solder joint cell geometry
(thickness) and (b) solder joint cell geometry (diameter)

encapsulation material (EMC), substrate, Cu pads, solder joints, solder mask, and PCB. The PBGA package type is BGA256, with dimensions of 21 mm × 21 mm and a ball pitch of 1 mm. The PCB dimensions are 300 mm × 180 mm × 2 mm. The dimensions of various components of the PBGA package solder joint unit are shown in Figure 5.4, including the thickness of the EMC (T_1), the thickness of the substrate (T_2), the thickness (T_3, T_4) and diameter (D_2, D_3) of the Cu pads, and the

Table 5.1   Geometry dimension of solder joint cell unit (mm)

Parameter	D1	D2	D3	H1	T1	T2	T3	T4	T5
Size	0.6	0.56	0.4	0.3	0.8	0.4	0.058	0.05	2

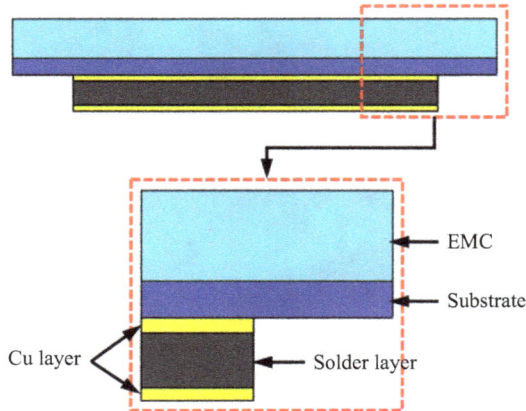

Figure 5.5   Simplified model of PBGA

height (H1) and diameter (D1) of the solder joints. The specific dimensions are listed in Table 5.1.

5.3.2   Finite-element model of PBGA packaging

As the sample contains 12 PBGA-packaged chips to shorten the finite-element analysis time and improve analysis efficiency, simplifications are made to the PBGA packaging sample model: (1) There is no residual stress or strain within the package body; (2) the influence of the solder mask on the calculation results is neglected; (3) the materials are ideally connected without voids or impurities; (4) all materials are isotropic; and (5) the Cu pads and solder joints of 11 PBGA-packaged chips are simplified into a uniform layer of medium, respectively. The simplified chip is shown in Figure 5.5, including EMC, substrate, Cu layer, and solder layer. The dimensions of the Cu layer on the device side are 15.56 mm × 15.56 mm × 0.058 mm, the solder layer dimensions are 15.56 mm × 15.56 mm × 0.3 mm, and the dimensions of the Cu layer on the PCB side are 15.56 mm × 15.56 mm × 0.05 mm.

After simplification, the finite-element model of the PBGA packaging sample includes: EMC, substrate, solder joints, PCB, Cu pads, Cu layer, and solder layer. Due to the symmetry of the PBGA packaging sample, only three 3D finite-element models need to be established for computation. The PBGA packaging finite-element model is created using ABAQUS. First, based on the aforementioned dimensional parameters, components for epoxy molding compounds (EMC),

Table 5.2   *Material properties of the PBGA*

Components	E (GPa)	ν	ρ (kg/m³)
EMC	25	0.26	2000
Substrate	22	0.38	4000
Cu pad	129	0.38	8940
Solder	36.2	0.38	8400
PCB	29.5	0.30	2315
Cu layer on the device side	129	0.38	2670.4
Soldering layer	36.2	0.38	2010.1
Cu layer on the PCB side	129	0.38	1503.0

substrate, Cu pads, solder joints, PCB, Cu layer, and solder layer are established. The PBGA packaging sample model is then assembled. Subsequently, material properties are assigned to the PBGA packaging sample model. Young's modulus (E), Poisson's ratio (ν), and density (ρ) of each part of the PBGA packaging sample are listed in Table 5.2. Finally, the PBGA packaging model is meshed to obtain the finite-element model, as shown in Figure 5.5.

Model 1 includes one non-simplified PBGA packaging chip (U1), while the other chips are simplified chip models. Model 1 contains 657,180 linear hexahedral elements (C3D8R) and 13,960 linear wedge elements (C3D6), totaling 751,962 nodes. Model 2 includes one non-simplified PBGA packaging chip (U2), while the other chips are simplified chip models. Model 2 contains 655,436 linear hexahedral elements (C3D8R) and 13,240 linear wedge elements (C3D6), totaling 750,370 nodes. Model 3 includes one non-simplified PBGA packaging chip (U3), while the other chips are simplified chip models. Model 3 contains 639,606 linear hexahedral elements (C3D8R) and 13,230 linear wedge elements (C3D6), totaling 747,322 nodes (Figure 5.6).

In the vibration test, the four corners of the sample are fixed to the test fixture with bolts, which are then secured to the vibration table. Therefore, in the finite-element analysis, full constraints are applied to the inner walls of the four bolt holes in the model, as shown in Figure 5.7.

5.3.3   *Finite-element model calibration*

The finite-element analysis model is calibrated by comparing it with the frequency sweep experimental results. In the sine sweep experiment, the sample is subjected to an acceleration amplitude of 0.5 G with sine vibrations in frequency ranges of 5–2000 and 50–500 Hz, with a sweep rate of 1 oct/min. The vibration direction is perpendicular to the front surface of the PCB and is defined as the Z-direction. During the finite-element analysis, both modal analysis and sine sweep vibration analysis are performed on the model. The sine sweep vibration is solved using the steady-state dynamics, modal analysis step. The load is consistent with the experiment, and the boundary conditions are shown in Figure 5.7.

The finite-element model's mass was compared with the sample's mass, and the comparison results are listed in Table 5.3. The mass of the model differs from

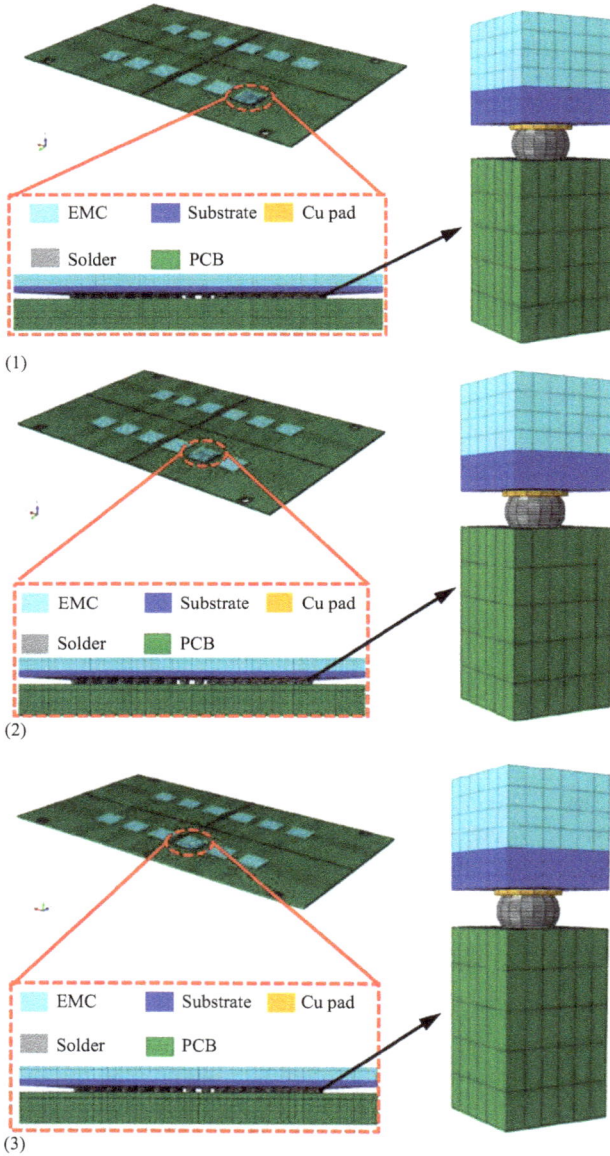

Figure 5.6   Finite-element model of PBGA specimen: (1) Model 1, (2) Model 2, and (3) Model 3

the mass of the sample by only 2.1%, which confirms the feasibility of the density parameters used in the finite-element model.

   The Z-directional acceleration response at the center of the sample obtained from the experiment is shown in Figure 5.8. The natural frequencies of the sample

Fully constrained inner wall of the bolt hole

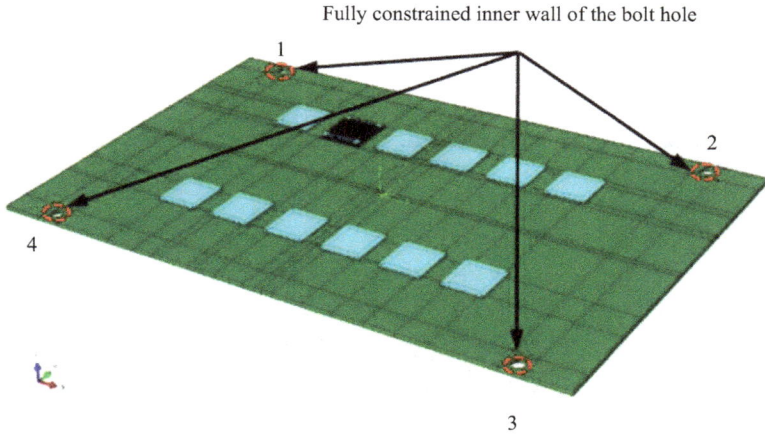

Figure 5.7   Boundary conditions of finite-element model

Table 5.3   Weight of finite-element model versus test vehicle

Model	Sample	Error
230 g	274.8 g	2.1%

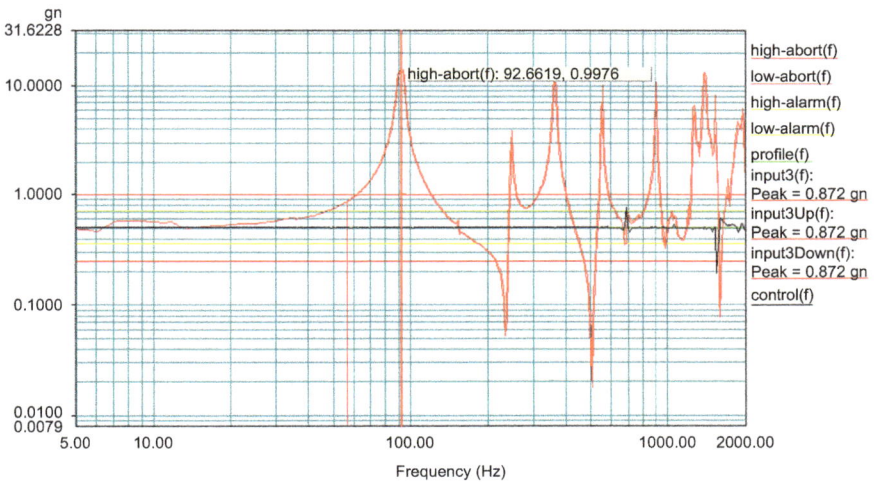

Figure 5.8   Acceleration response of test vehicle

obtained from the frequency sweep experiment are listed in Table 5.4. The first three mode shapes of the three models obtained from the finite-element model analysis are shown in Figure 5.9, and the natural frequencies of the models are listed in Table 5.5.

Table 5.4   Natural frequency of test vehicle

Natural frequency	First-order	Second-order	Third-order
Value	92.67 Hz	233.98 Hz	248.11 Hz

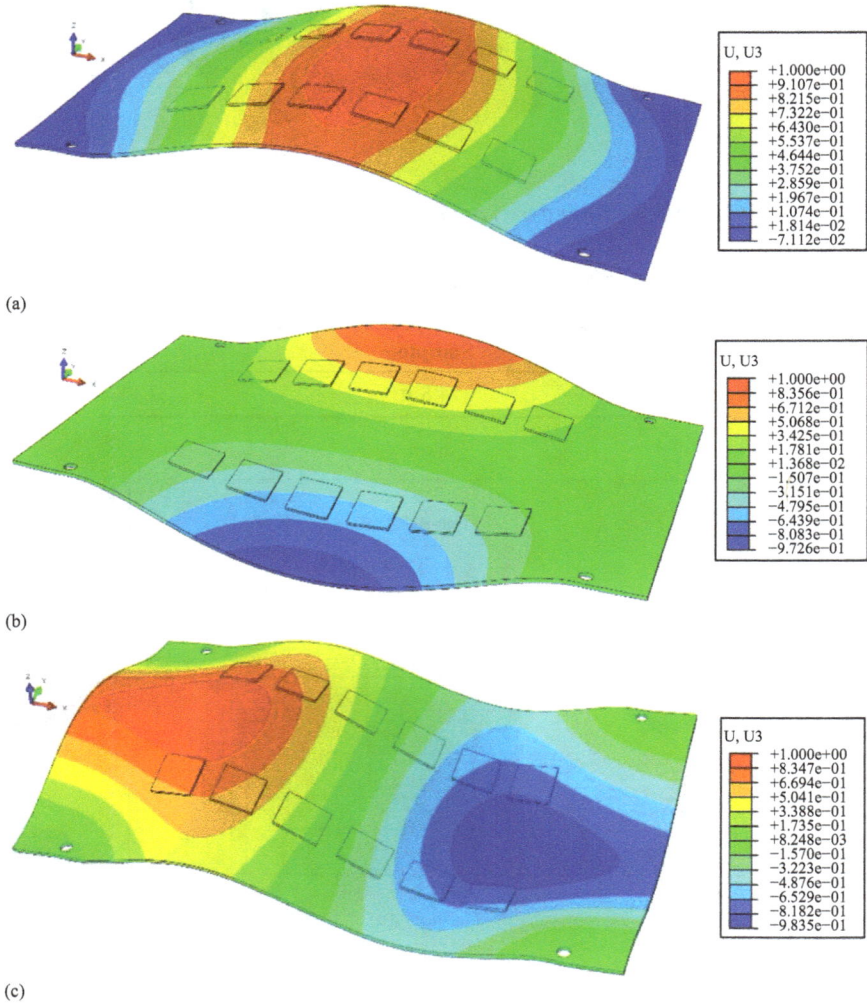

(a)

(b)

(c)

Figure 5.9   The first three modal shape of finite-element model: (a) first modal shape, (b) second modal shape, and (c) third modal shape

Table 5.5   Natural frequency of finite-element model

Natural frequency	Model 1	Model 2	Model 2
First-order (Hz)	93.42	93.49	93.50
Second-order (Hz)	185.14	186.34	188.49
Third-order (Hz)	232.09	232.27	232.73

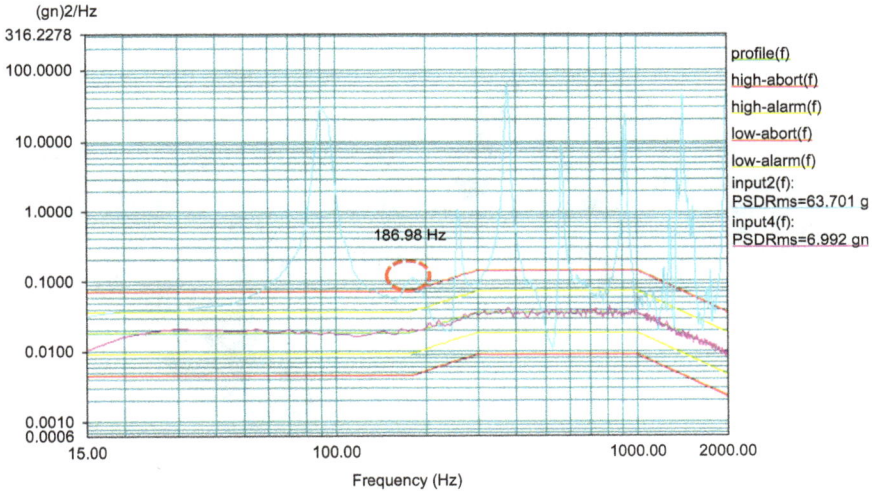

Figure 5.10   Acceleration response of the PCB center for random vibration at the Z direction

Due to a significant discrepancy between the second-order natural frequency obtained from the experiment and that from the finite-element analysis, with a maximum error of 20.9%, the second-order natural frequency of the sample was remeasured using random vibration sweep testing, resulting in 186.98 Hz. The Z-directional acceleration response at the center of the PCB from the random vibration sweep test is shown in Figure 5.10. The comparison of the sample's natural frequencies with those of the model is shown in Table 5.6. The maximum error between the first three natural frequencies of the three models and the experimental results is 0.99%, demonstrating the accuracy of the finite-element model.

During the harmonic sweep vibration finite-element analysis, the response of the harmonic sweep vibration is related to damping. To determine the equivalent damping of the sample, multiple finite-element analyses of the sine sweep with a 0.5 G load were conducted and compared with the experimental acceleration and displacement responses. In the sine sweep analysis, a base excitation method was used, applying Z-directional sine excitation with an acceleration amplitude of 0.5 G to the bolt hole walls at the four corners of the sample, with the frequency sweep

Table 5.6   Natural frequency of finite-element model versus test vehicle

Natural frequency	Model 1	Model 2	Model 2	Experiment	Maximum error (%)
First-order	93.42	93.49	93.50	92.67	0.90
Second-order	185.14	186.34	188.49	186.98	0.99
Third-order	232.09	232.27	232.73	233.98	0.81

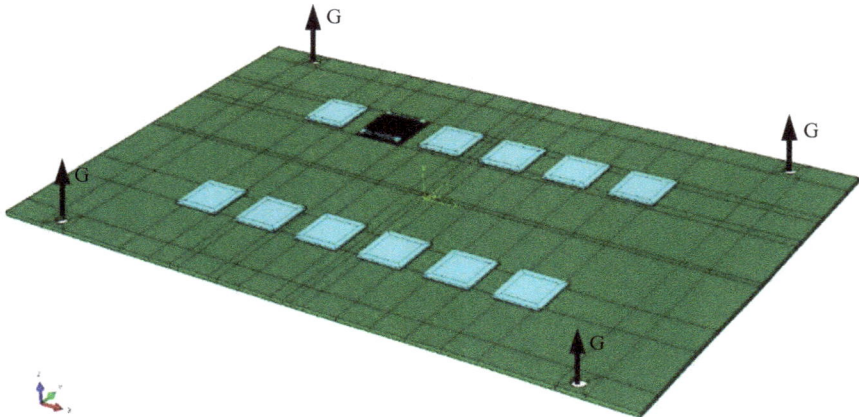

Figure 5.11   Load application technique

range consistent with the experimental conditions. The loading method is shown in Figure 5.11.

It was ultimately determined that when the modal damping is set to 0.023, the finite-element analysis results for the Z-directional acceleration response and displacement response at the center of the PCB at the first natural frequency closely match the experimental results. Figure 5.12 shows the Z-directional acceleration response at the PCB center obtained from finite-element analysis. The acceleration response at the first natural frequency is listed in Table 5.7, with a 7.2% error between the finite-element analysis results and experimental results.

Figure 5.13 presents the peak-to-peak displacement response at the PCB center in the Z-direction with an input acceleration amplitude of 0.5 G and a frequency sweep range of 50–500 Hz. Figure 5.14 shows the Z-directional displacement amplitude response at the PCB center obtained from finite-element analysis. The displacement amplitude response at the first natural frequency is listed in Table 5.8, with an 8.7% error between the finite-element analysis and experimental results. The comparison of acceleration and displacement responses demonstrates the feasibility of the damping value used in the finite-element model, thus validating the correctness and feasibility of the finite-element model.

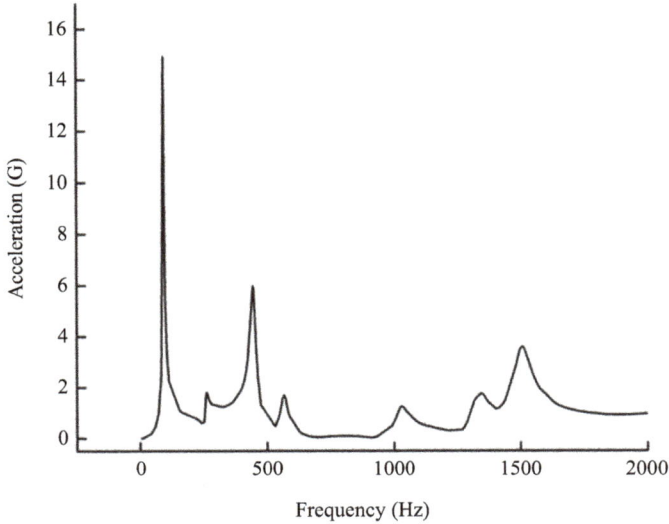

Figure 5.12 Acceleration response of PCB center at the Z direction

Table 5.7 Acceleration response of finite-element model versus test vehicle

Acceleration	Experiment	Finite-element analysis	Error
First-order response	13.9 G	14.9 G	7.2%

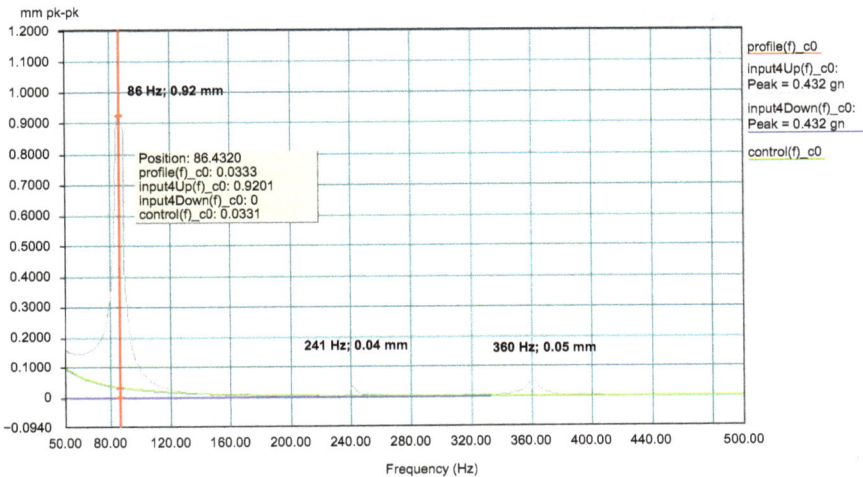

Figure 5.13 Displacement response of the PCB center for the experiment at the Z direction

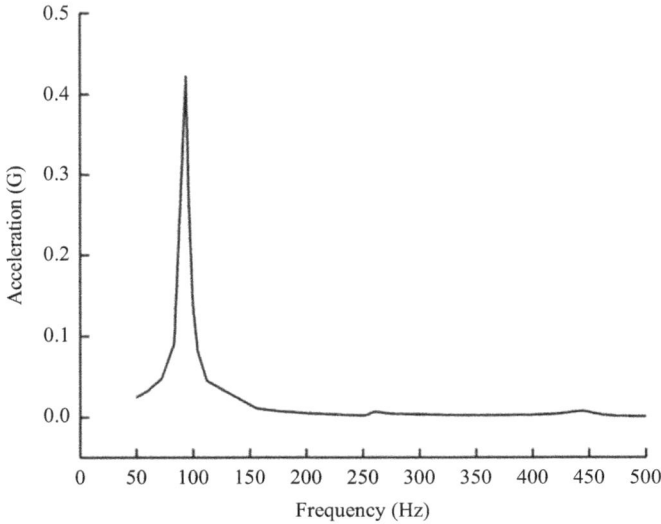

Figure 5.14   Displacement response of the PCB center for simulation at the Z direction

Table 5.8   Acceleration response of finite-element model versus test vehicle

Acceleration	Experiment	Finite-element analysis	Error
First-order response	0.46 mm	0.42 mm	8.7%

5.4   Stress analysis of PBGA package solder joints under harmonic vibration loads

As shown in Figure 5.6, three-dimensional finite-element models of the three PBGA package samples were established. Based on the conditions of the harmonic sweep vibration experiment, the inner walls of the bolt holes on the PCB were fully constrained. Base excitation was used for loading, applying sinusoidal vibration loads with acceleration amplitudes of 5, 7, and 10 G within a sweep range of 87–97 Hz to the constrained boundary, with the vibration direction perpendicular to the upper surface of the PCB. The finite-element analysis was performed using the steady-state dynamics, modal solver in ABAQUS.

For ease of result analysis, considering the symmetry of the PBGA package samples, the PBGA package chips were divided into three groups based on their distance from the center of the PCB: group 1, group 2, and group 3, as listed in Table 5.9. The PBGA package solder joints were numbered as shown in Figure 5.15. Harmonic sweep analysis was conducted on the three PBGA package sample models, and the von Mises stress distribution for the representative 10 G

Table 5.9   Groups of PBGA components

Group	Group 1	Group 2	Group 3
Chip number	U3 U4 U9 U10	U2 U5 U8 U11	U1 U6 U7 U12

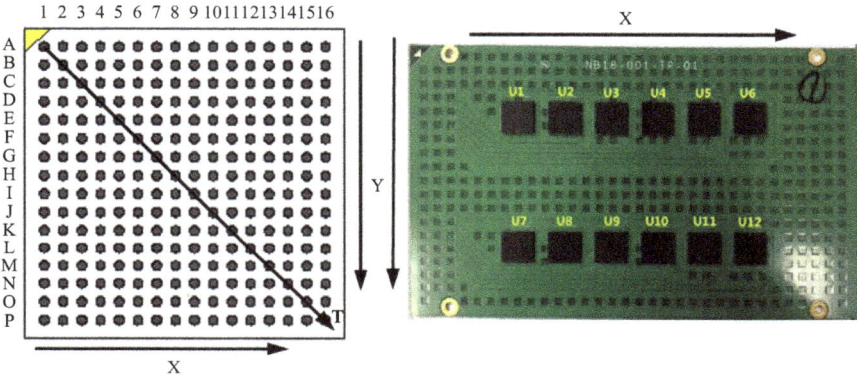

Figure 5.15   Number and orientation of solder joint arrays

load condition across each group of PBGA package solder joints is shown in Figure 5.16. It can be observed that each solder joint array exhibits higher stress in the corner regions and lower stress in the central regions. For the three load conditions, the maximum stress values for each group of PBGA package solder joints are listed in Table 5.10. It can be seen from the table that, under the same excitation load, the stress is highest for group 1, followed by group 2, and lowest for group 3.

Figure 5.17 illustrates the variation of maximum stress in PBGA package solder joints along the diagonal *T* direction, with the position and orientation of the diagonal indicated in Figure 5.2. The analysis reveals that under harmonic sweep vibration loads, the maximum stress in PBGA package solder joints increases progressively as the chip position approaches the center of the PCB. Under a 5 G load, the maximum stress in the solder joints of group 1 is 191.96 MPa, whereas for group 3, it is 77.09 MPa. The maximum stress in group 1 solder joints is 2.49 times that in group 3 solder joints. Under a 7 G load, the maximum stress in the solder joints of group 1 is 268.34 MPa, while in group 3, it is 107.93 MPa. The maximum stress in group 1 solder joints is again 2.49 times that in group 3 solder joints. Under a 10 G load, the maximum stress in the solder joints of group 1 is 383.93 MPa, compared to 154.19 MPa in group 3. The maximum stress in group 1 solder joints is 2.49 times that in group 3 solder joints. These results indicate that solder joints located closer to the PCB center are more susceptible to failure compared to those positioned further away. For the same input conditions, the maximum stress

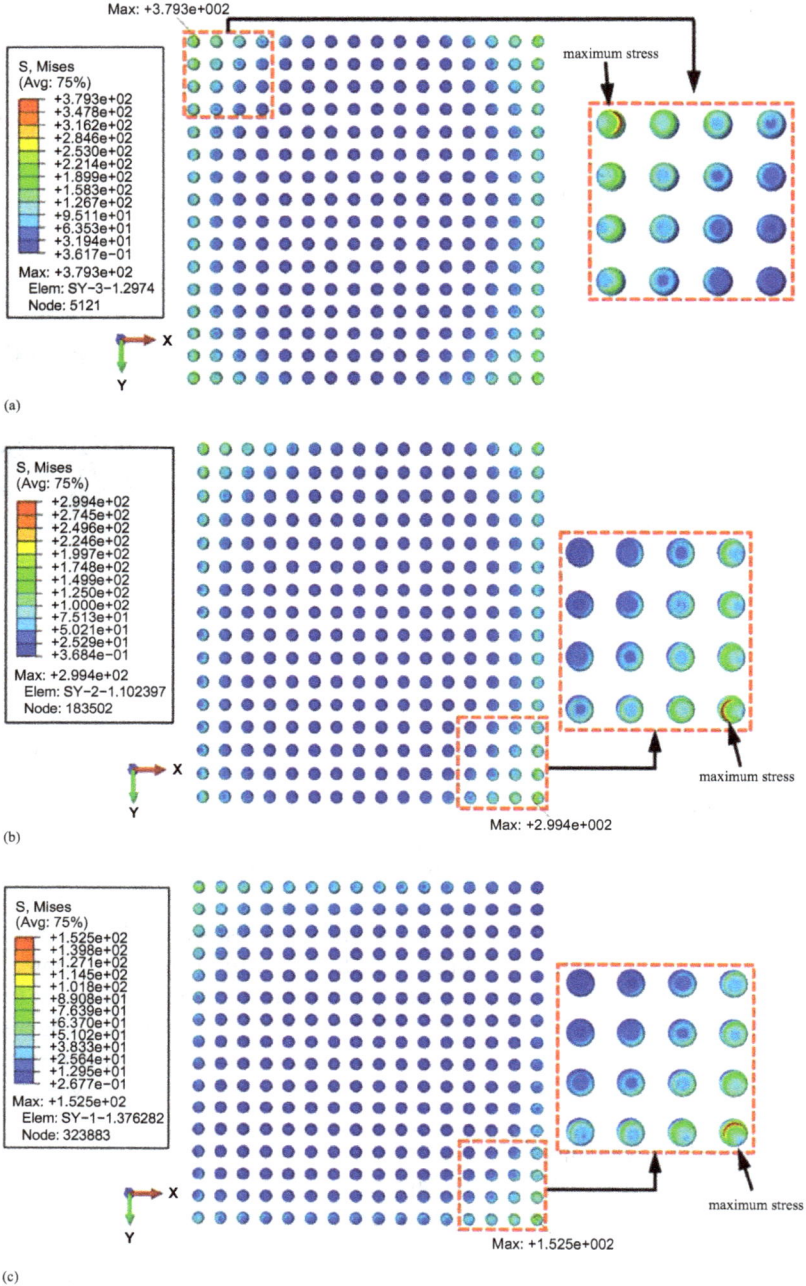

Figure 5.16   Stress contour of solder joints for three PBGA groups under harmonic sweep vibration: (a) stress contour of group 1 solder joints, (b) stress contour of group 2 solder joints, and (c) stress contour of group 3 solder joints

Table 5.10   The maximum stress of solder joints for three PBGA groups

Loading (G)	Group 1 (MPa)	Group 2 (MPa)	Group 3 (MPa)
5	191.96	149.08	149.08
7	268.34	208.71	208.71
10	383.93	298.16	298.16

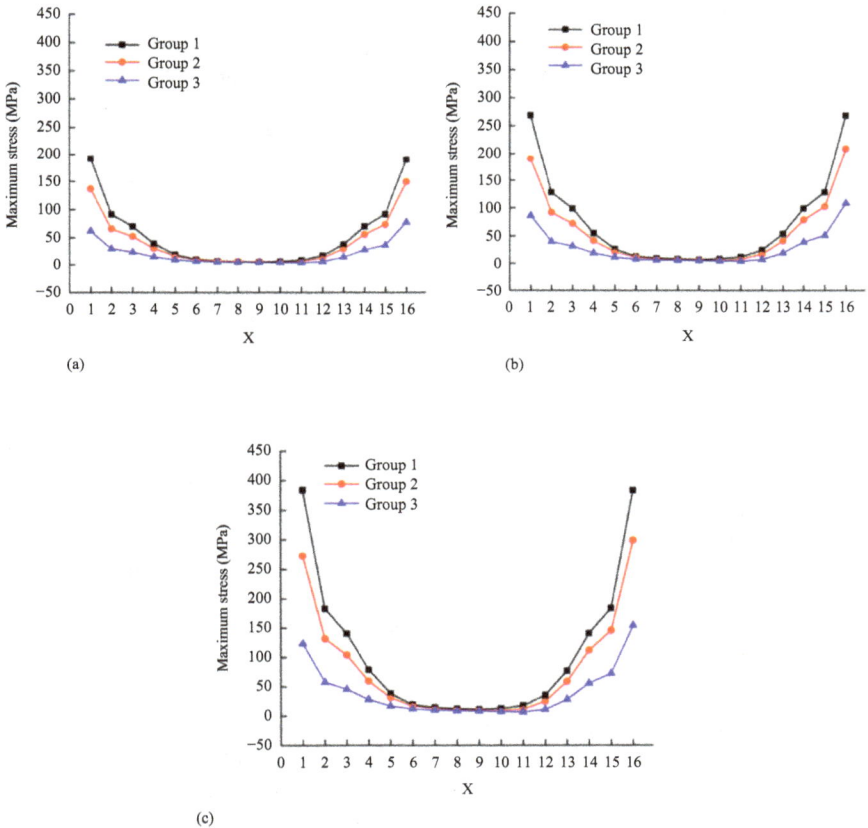

(a)

(b)

(c)

Figure 5.17   Maximum stress in solder joints along the T direction for three groups: (a) Maximum stress in solder joints along the T direction under the load condition of 5 G, (b) maximum stress in solder joints along the T direction under the load condition of 7 G, and (c) maximum stress in solder joints along the T direction under the load condition of 10 G

in solder joints nearer to the PCB center (group 1) is consistently 2.49 times greater than that in solder joints located farther from the PCB center (group 3).

The observed results can be attributed to the fact that during the harmonic sweep vibration process, the input excitation frequency ranges from 87 to 97 Hz, which is centered around the first natural frequency of the PBGA package sample (92.67 Hz). Under this excitation, the mode shape of the sample closely resembles the first mode shape (as shown in Figure 5.9). Due to the repeated bending of the PCB, alternating stress is generated in the solder joints connected to it. The maximum bending deformation occurs at the position of group 1 on the PCB, resulting in the highest alternating stress in the solder joints, followed by group 2, and then group 3 with the lowest stress.

Additionally, as shown in Figure 5.17, the position of the chip has a minimal effect on the maximum stress in the solder joints located in the central region of the chip; instead, the chip position primarily influences the maximum stress in the solder joints located at the periphery of the chip. Figure 5.18 presents the variation curve of the maximum stress in the PBGA package solder joints along the Y direction of the 16th column. From the figure, it can be observed that the maximum stress variation curve for group 3 solder joints differs from those for groups 1 and 2.

The maximum stress in the solder joints at the positions of groups 1 and 2 exhibits a pattern where the stress is higher at the outer solder joints and lower at the central solder joints. However, for group 3 solder joints, the stress distribution shows that the stress is lower on one side of solder joint (16, A) and higher on the side of solder joint (16, P), with the maximum stress increasing continuously from solder joint (16, B) to solder joint (16, P).

Under harmonic sweep vibration loads, the stress contour maps for representative solder joint arrays are shown in Figure 5.19. At a 10 G load, the stress in the PBGA package solder joints demonstrates that the stress is higher at the corner solder joints and lower at the central solder joints. The maximum stress in the entire package solder joint array is observed at the corner solder joints of the array. For the three different load levels, the maximum stress for group 1 solder joints occurs at solder joint (1, A), while for both group 2 and group 3 solder joints, the maximum stress is observed at solder joint (16, P).

Figure 5.20 illustrates the variation of maximum stress in PBGA package solder joints along the diagonal T direction. The maximum stress in the solder joints shows a decreasing trend from solder joint (1, A) to solder joint (6, F). From solder joint (6, F) to solder joint (11, K), the maximum stress remains relatively stable. From solder joint (11, K) to solder joint (16, P), the maximum stress exhibits an increasing trend. Additionally, Figure 5.20 indicates that, across different vibration magnitudes, there is minimal variation in the maximum stress of solder joints in the central region of the PBGA package, while significant differences are observed in the maximum stress of solder joints in the peripheral region. For group 1 packages, the maximum stress at solder joint (1, A) under a 5 G load is 191.96 MPa, whereas under a 10 G load, it increases to 383.93 MPa, which is 2.00 times the maximum stress at 5 G. For group 2 packages, the maximum stress at solder joint (1, A) under a 5 G load is 149.08 MPa, and under a 10 G load, it increases to

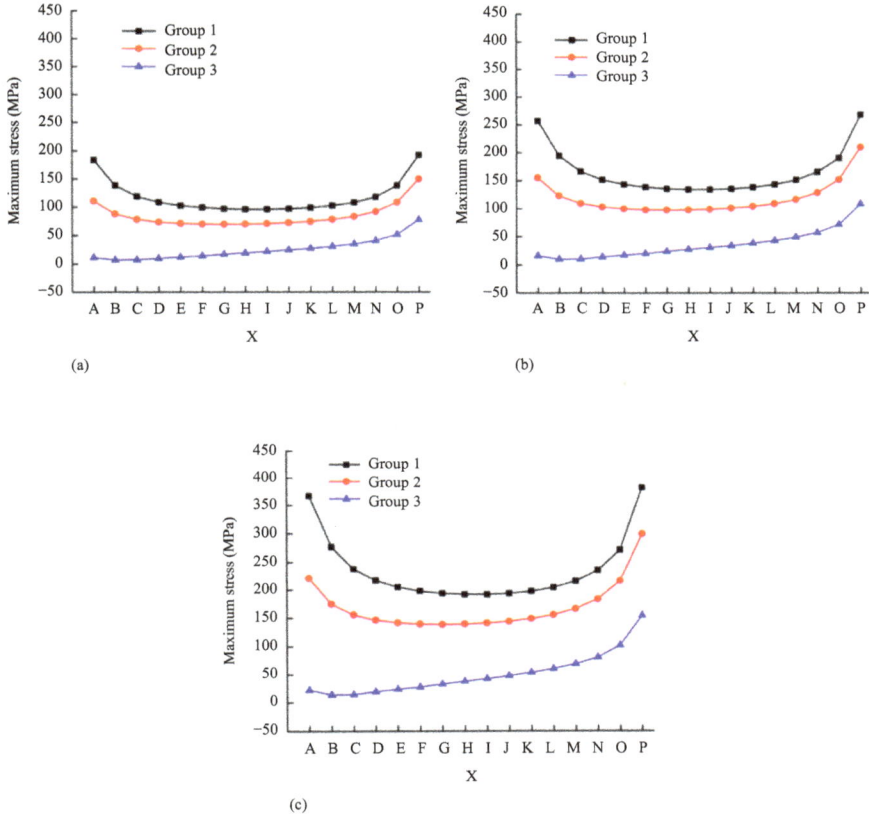

Figure 5.18   *Maximum stress in solder joints at column 16 for three groups:*
(a) maximum stress in solder joints at column 16 under the load
condition of 5 G, (b) maximum stress in solder joints at column 16
under the load condition of 7 G, and (c) maximum stress in solder
joints at column 16 under the load condition of 10 G

298.16 MPa, which is also 2.00 times the maximum stress at 5 G. For group 3 packages, the maximum stress at solder joint (1, A) under a 5 G load is 77.09 MPa, and under a 10 G load, it rises to 154.19 MPa, again 2.00 times the maximum stress at 5 G. It can be observed that for solder joints at the same position in the PBGA package, the maximum stress consistently doubles when the harmonic sweep vibration input increases from 5 to 10 G.

Due to the difference in bending stiffness between the PCB and the device, the outermost solder joints experience higher compressive stress when the PCB bends downward and higher tensile stress when the PCB bends upward. During harmonic sweep vibration, the PCB undergoes cyclic bending due to mechanical vibrations, causing the solder joints to endure both mechanical shock and alternating tensile-

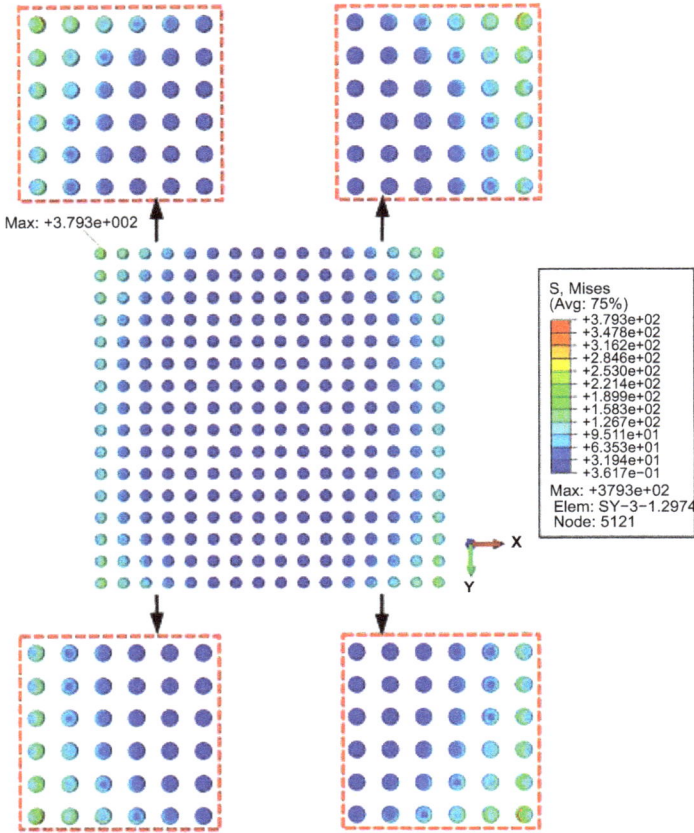

Figure 5.19   Stress contour of PBGA solder joints under harmonic sweep vibration

compressive stress from the PCB's cyclic bending, leading to solder joint failure. As the amplitude of acceleration increases, the bending degree of the PCB also increases, resulting in greater stress in the solder joints.

The stress analysis of the PBGA package solder joint array under harmonic sweep vibration loads reveals that the maximum stress occurs at the corner solder joints, making them the most susceptible to failure. Moreover, the locations of maximum stress in the solder joints need to be identified. Figures 5.21–5.23 present stress contour maps of critical solder joints at different positions under 5, 7, and 10 G loads, respectively. These contour maps show that the maximum stress in these critical solder joints is consistently located on the side of the solder joint closest to the device and along the edge of the solder joint. Figure 5.24 shows the stress response curves of key solder joints under a 5 G load, demonstrating that stress

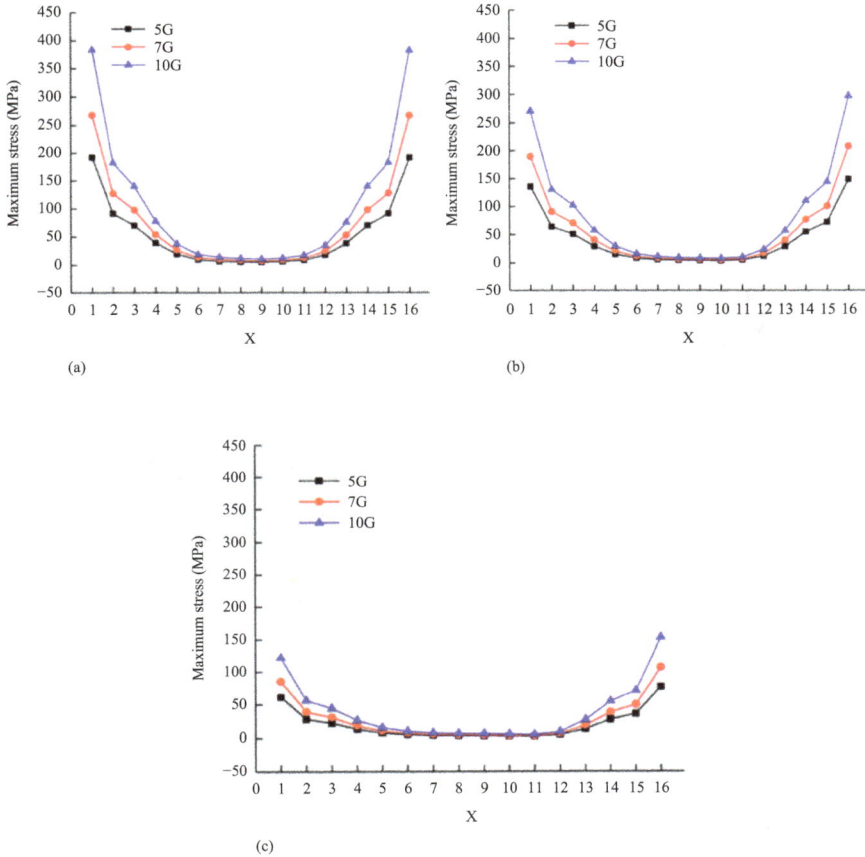

Figure 5.20 Maximum stress in PBGA solder joints along the T direction for three groups: (a) maximum stress in solder joints along the T direction for group 1, (b) maximum stress in solder joints along the T direction for group 2, and (c) maximum stress in solder joints along the T direction for group 3

increases with frequency up to the natural frequency, where resonance occurs, and stress reaches its peak. Beyond this point, as the frequency continues to increase, the stress in the solder joints begins to decrease.

Figure 5.25 presents the cross-sectional stress contour map of representative critical solder joints in PBGA packages under a 7 G load. The stress contour map indicates higher stress at the edge of the solder joint closest to the device, suggesting that fatigue cracks are likely to initiate at the edge of the solder joint under this stress. During vibration, these cracks are expected to propagate along the weak interface of the solder joint, eventually leading to failure. Figure 5.26 shows the

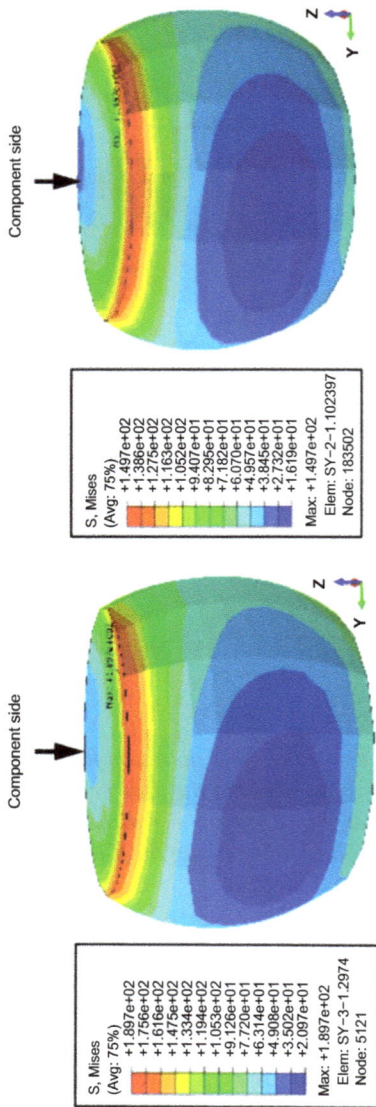

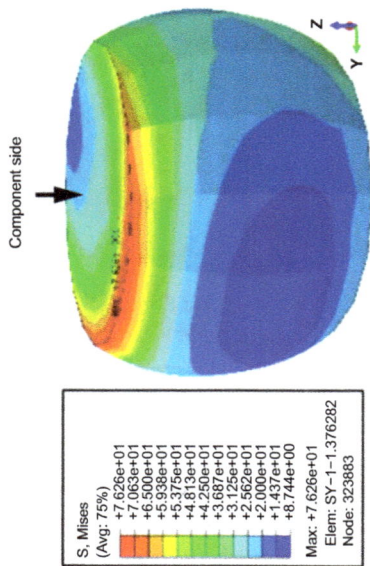

Figure 5.21 Stress contour of critical solder joints under the load condition of 5 G: (a) group 1, (b) group 2, and (c) group 3

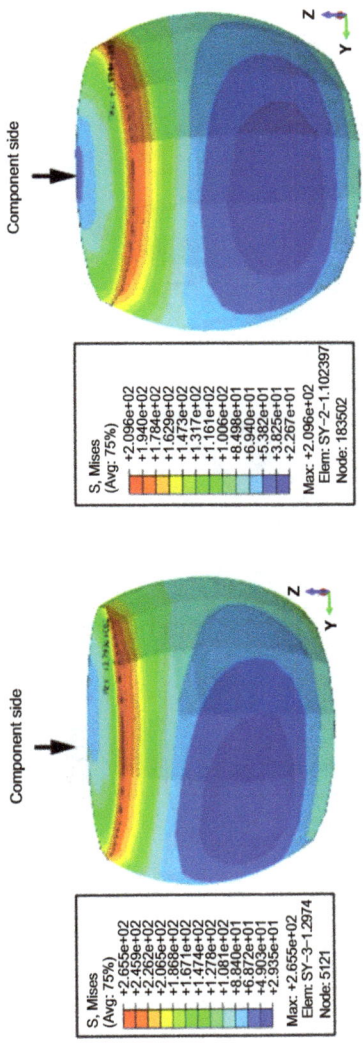

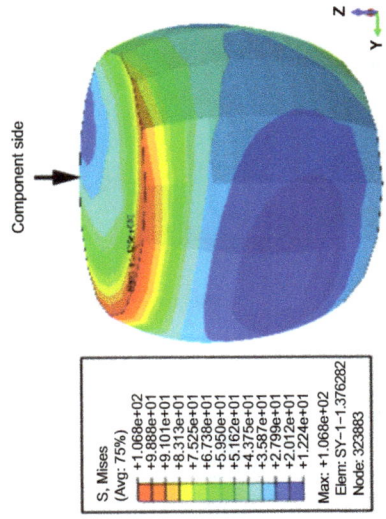

Figure 5.22 Stress contour of critical solder joints under the load condition of 7 G: (a) group 1, (b) group 2, and (c) group 3

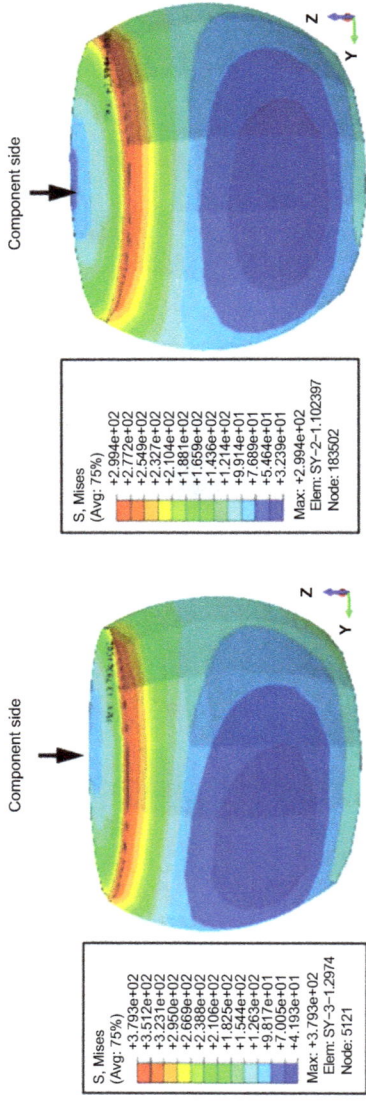

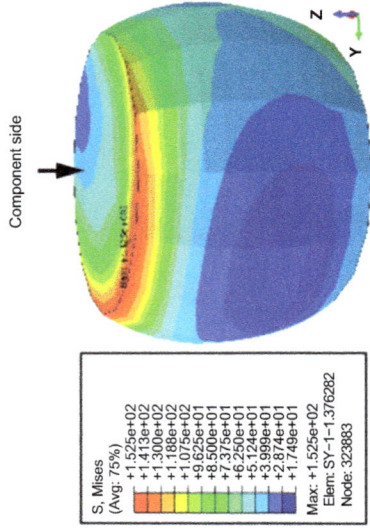

Figure 5.23   Stress contour of critical solder joints under the load condition of 10 G: (a) group 1, (b) group 2, and (c) group 3

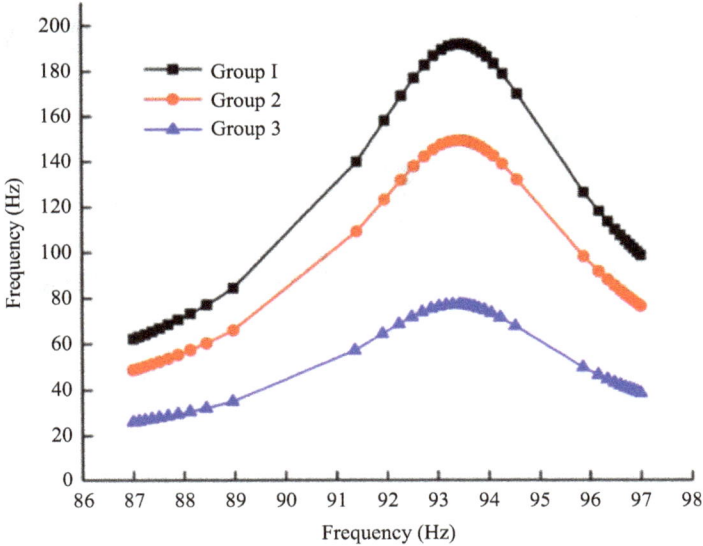

Figure 5.24   Stress response of critical solder joints under the load condition of 5 G

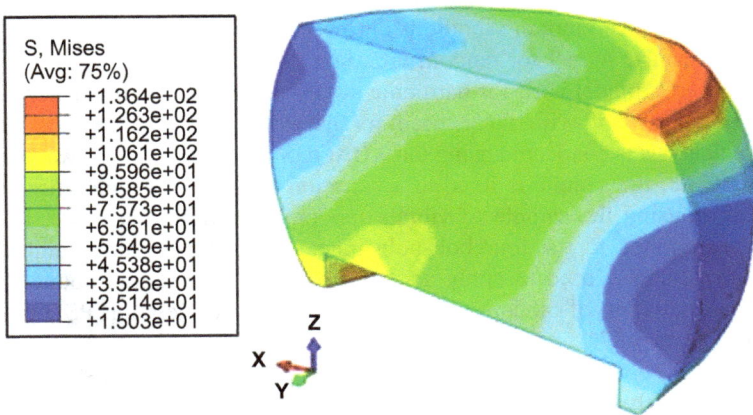

Figure 5.25   Cross-section of critical solder joint stress contour under harmonic sweep vibration

morphology and crack patterns of failed solder joints at the corner position under a 7 G load. Fatigue cracks that penetrate the solder joints are observed on the side closest to the device, and these findings are consistent with the results from the aforementioned finite-element analysis.

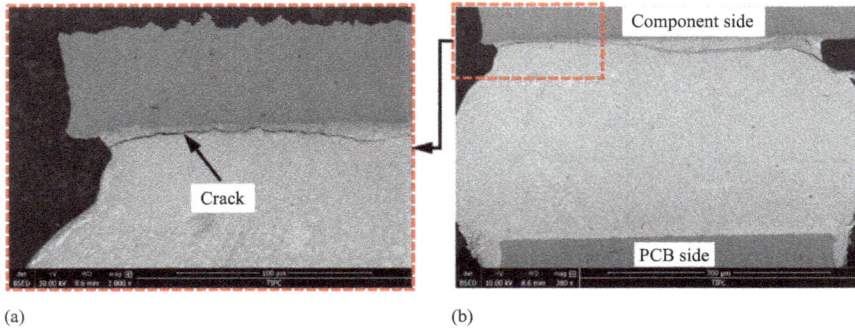

(a)   (b)

*Figure 5.26   SEM image of failure solder joint: (a) crack morphology (1000×)
and (b) morphology of the failure solder joint (380×)*

5.5   Stress analysis of PBGA package solder joints under random vibration loads

Many vibration phenomena in nature and engineering are stochastic in nature. When the vibration conditions of a system cannot be described by a specific function and past data cannot accurately predict future vibrations, the vibrations are considered random. Such non-reproducible mechanical or structural systems are studied using probabilistic statistical methods [5–7]. The PBGA package devices primarily experience random vibrations during their service life.

The geometric model for random vibration of the PBGA package is the same as the geometric model used for harmonic sweep vibration, and the material parameters are consistent with those used for harmonic sweep vibration. Random vibration analysis is conducted using the ABAQUS finite-element software. During the experiment, the sample is fixed to a vibration table, which undergoes random vibrations, causing the sample to vibrate randomly as well. Therefore, full constraints are applied to the four bolt-hole inner walls of the model, and a basic excitation method is used to apply acceleration PSD to the fixed boundaries for numerical simulation of random vibrations. The input acceleration PSD profile is shown in Figure 5.27.

The ABAQUS random vibration analysis is performed in two steps: the first step is modal analysis, and the second step is random response analysis. According to the random vibration experimental conditions, the analysis frequency is set from 15 to 2000 Hz, the damping coefficient is set to 0.023, and acceleration loads are applied as described above. Finally, the job is submitted for analysis. In this study, three levels of acceleration PSD are applied to the model based on JESD22-B103B.01, with root mean square (RMS) values (G_{RMS}) of 7, 11, and 18 G.

In random vibration, the RMS of the response is used to measure the fatigue strength of structures. ABAQUS provides the RMS value of the response, and in this section, von Mises stress RMS is used to measure the stress at solder joints. Random vibration analysis is performed on three PBGA package sample models,

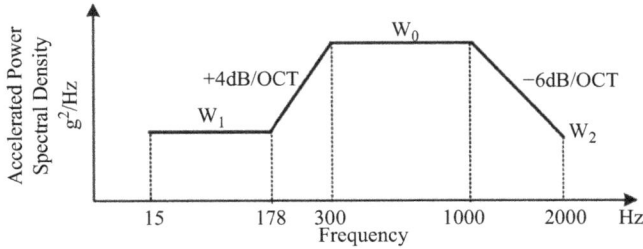

Figure 5.27   Accelerated PSD profile for random vibration

and the stress at solder joint arrays is extracted. A representative distribution of PBGA package solder joints is shown in Figure 5.28. Under three different load magnitudes, the maximum σRMS of each group of PBGA package solder joints is listed in Table 5.11. From the table, it can be seen that under the same excitation load, the stress is highest in the first group of solder joints, followed by the second group, and lowest in the third group.

Figure 5.29 shows the maxim um σ_{RSM} variation curve for solder joints along the diagonal T direction of the PBGA package. Under random vibration loads, the maximum stress at the PBGA package solder joints increases as the chip position approaches the PCB center. When the σ_{RSM} is 7 G, the maximum σ_{RSM} of the first group of solder joints is 13.60 MPa, while the maximum σ_{RSM} of the third group is 6.09 MPa. The maximum σ_{RSM} of the first group is 2.23 times that of the third group. When the G_{RMS} is 11 G, the maximum σ_{RSM} of the first group is 21.56 MPa, and the maximum σ_{RSM} of the third group is 9.63 MPa. The maximum σ_{RSM} of the first group is 2.24 times that of the third group. When the σ_{RSM} is 18 G, the maximum σ_{RSM} of the first group is 35.01 MPa, and the maximum σ_{RSM} of the third group is 15.67 MPa. The maximum σ_{RSM} of the first group is 2.23 times that of the third group.

It can be observed that under the same load, the PBGA package solder joints near the PCB center are more prone to failure compared to those farther from the PCB center, with the maximum stress of solder joints near the PCB center (first group) being approximately 2.23 times that of the solder joints farther from the PCB center (third group). Additionally, as shown in Figure 5.29, similar to harmonic sweep vibration, the position of the chip relative to the PCB has a smaller effect on the stress of the solder joints in the central area of the PBGA package during random vibration, but a greater effect on the stress of the solder joints in the peripheral areas of the PBGA package. Figure 5.30 shows the variation curve of the maximum σ_{RSM} for solder joints along the Y direction of the 16th column. From Figure 5.30, it can be seen that the maximum σ_{RSM} of the three groups of solder joints exhibits a pattern where solder joints at the edges have higher stress, while solder joints in the central area have lower stress.

Under random vibration loads, the stress distribution maps of representative solder joint arrays are shown in Figure 5.31. When the input G_{RMS} is 18 G, the

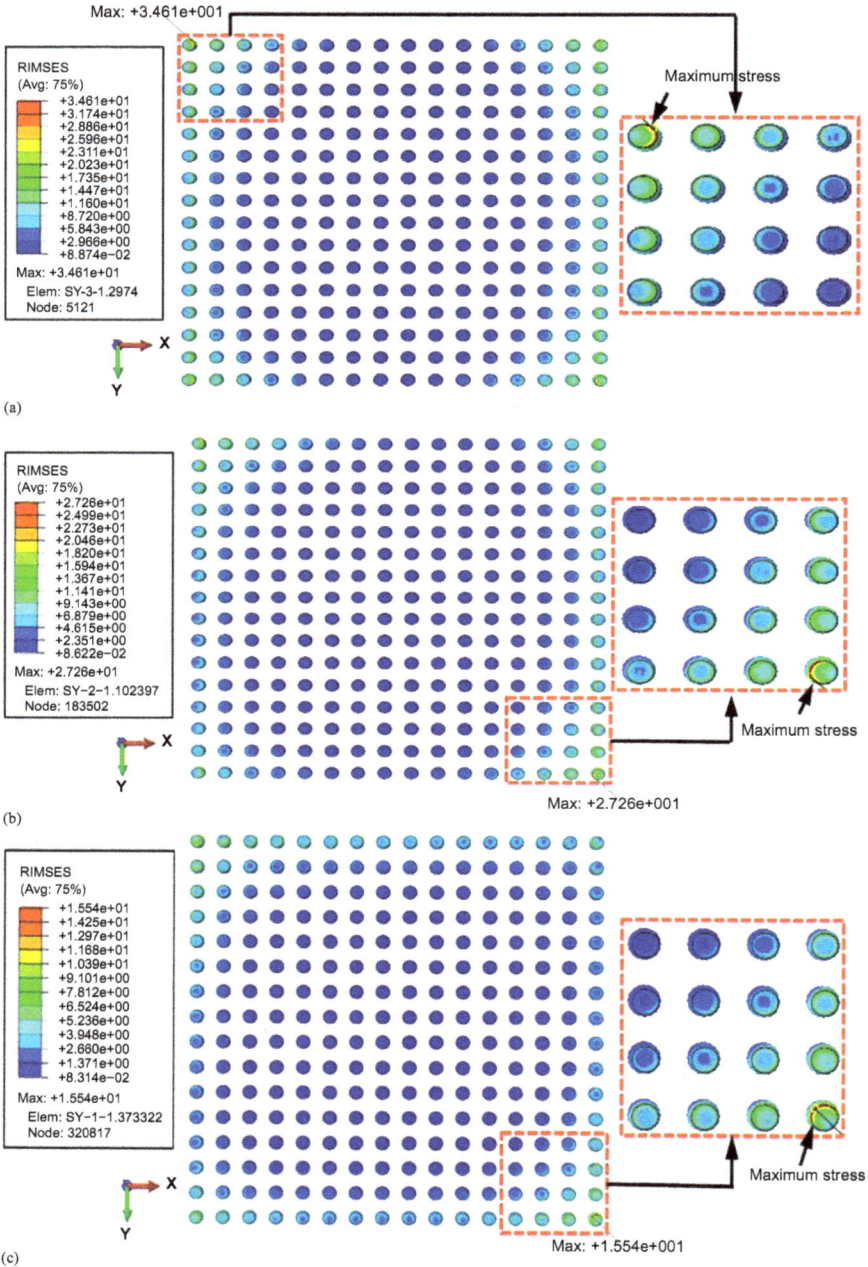

Figure 5.28   Stress contour of solder joints for three PBGA groups during random vibration: (a) stress contour of group 1 solder joints, (b) stress contour of group 1 solder joints, and (c) stress contour of group 1 solder joints

Table 5.11   The maximum stress in solder joints for three PBGA groups

G_{RMS} (G)	Group 1 (MPa)	Group 2 (MPa)	Group 3 (MPa)
7	13.60	13.60	13.60
11	21.56	21.56	21.56
18	35.01	35.01	35.01

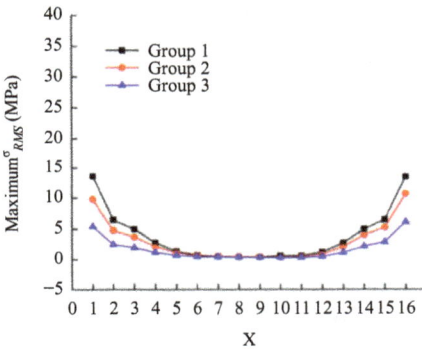

(a)

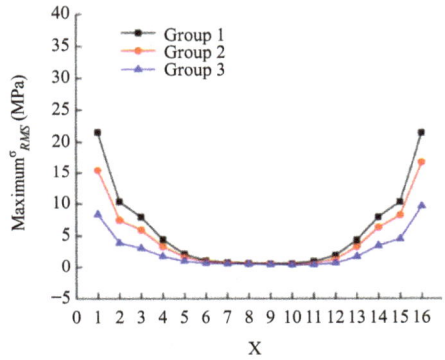

(b)

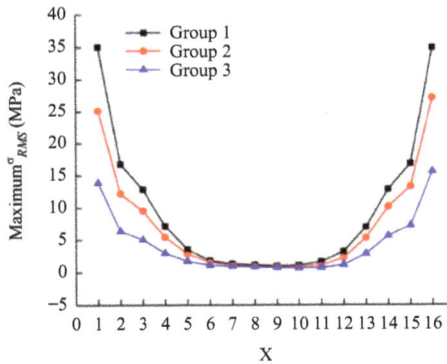

(c)

Figure 5.29   Maximum stress in solder joints at column 16 under random vibration for three groups: (a) maximum stress in solder joints at column 16 under the load condition of G_{RMS}=7 G, (b) maximum stress in solder joints at column 16 under the load condition of G_{RMS}=10 G, and (c) maximum stress in solder joints at column 16 under the load condition of G_{RMS}=18 G

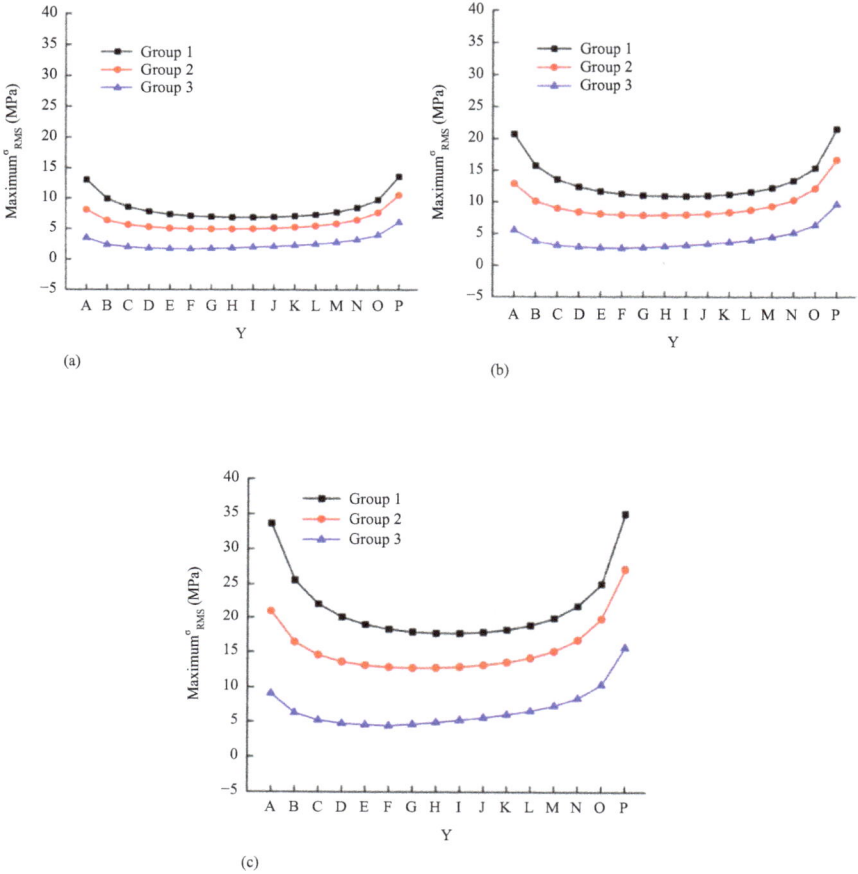

*Figure 5.30   Maximum stress in solder joints at column 16 under random
vibration for three groups: (a) maximum stress in solder joints at
column 16 under the load condition of G_{RMS}=7 G, (b) maximum
stress in solder joints at column 16 under the load condition of
G_{RMS}=10 G, and (c) maximum stress in solder joints at column 16
under the load condition of G_{RMS}=18 G*

stress in the PBGA package solder joints shows that the stress is higher at the
corner region and lower at the central region. The maximum stress in the entire
solder joint array occurs at the corner of the solder joint array. Among the three
load cases, the maximum stress for group 1 solder joints is observed at the solder
joint (1, A), while for groups 2 and 3, the maximum stress is observed at the solder
joint (16, P).

Figure 5.32 shows the maximum stress variation curve of PBGA package
solder joints along the diagonal T direction. The maximum stress of the solder

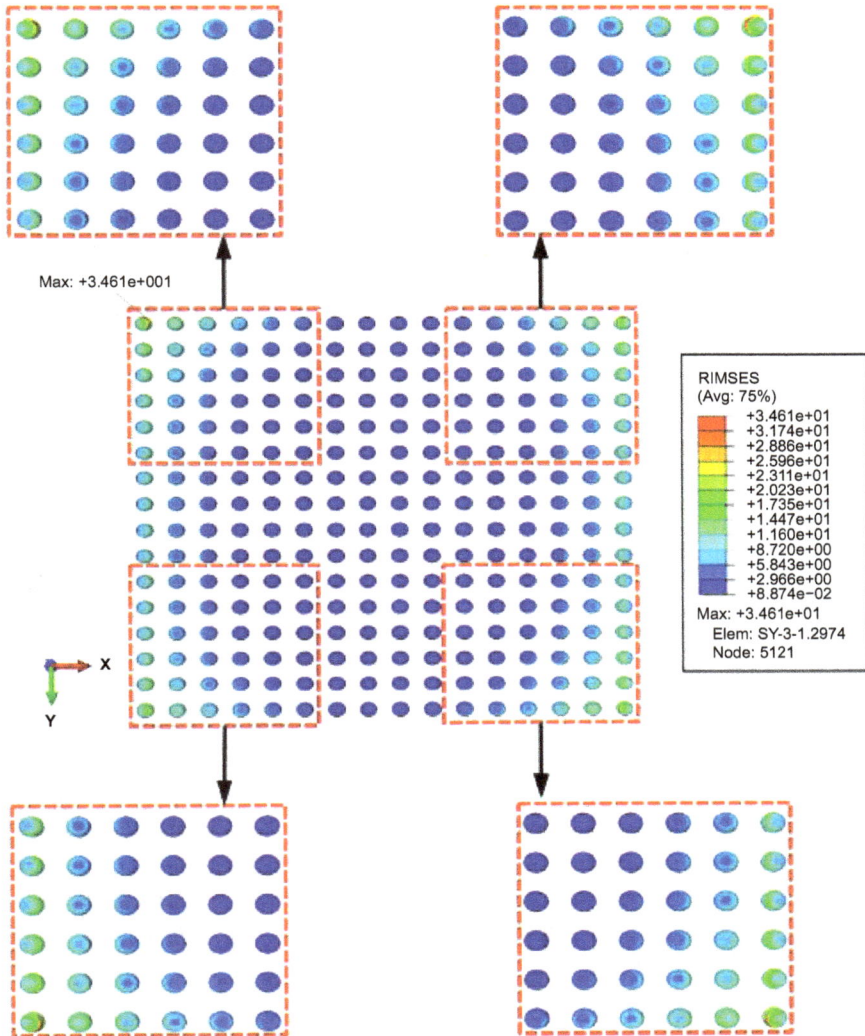

Figure 5.31   Stress contour of PBGA solder joints under random vibration

joints decreases from solder joint (1, A) to solder joint (6, F); it remains stable from solder joint (6, F) to solder joint (11, K) and increases from solder joint (11, K) to solder joint (16, P). This indicates that under random vibration loads, PBGA package solder joint failures primarily occur in the corner regions of the solder joint array. This is similar to the stress response observed in PBGA package solder joints under harmonic sweep vibration loads, as discussed earlier. Additionally, Figure 5.32 reveals that different vibration levels have little impact on the

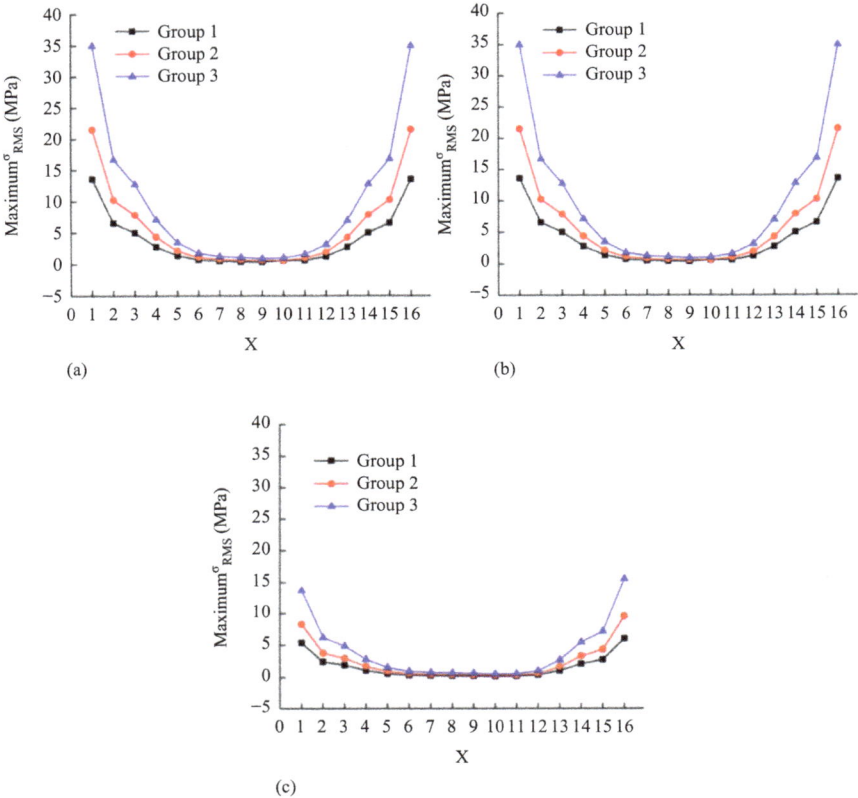

(a)

(b)

(c)

Figure 5.32   Maximum stress in PBGA solder joints along the T direction for three groups: (a) maximum stress in solder joints along the T direction for group 1, (b) maximum stress in solder joints along the T direction for group 2, and (c) maximum stress in solder joints along the T direction for group 3

maximum stress of solder joints in the central region of the package but significantly affect the maximum stress of solder joints in the peripheral region of the package.

For group 1 chips, the maximum σ_{RSM} of solder joints is 13.60 MPa at a σ_{RSM} of 7 G and 35.01 MPa at a σ_{RSM} of 18 G, which is 2.57 times the maximum σ_{RSM} at 7 G. For group 2 chips, the maximum σ_{RSM} is 10.63 MPa at 7 G and 27.15 MPa at 18 G, which is 2.55 times the maximum σ_{RSM} at 7 G. For group 3 chips, the maximum σ_{RSM} is 6.09 MPa at 7 G and 15.67 MPa at 18 G, which is 2.57 times the maximum σ_{RSM} at 7 G. It can be observed that for solder joints at the same location in the PBGA package, the maximum σ_{RSM} of the solder joints increases by nearly 2.57 times when the random vibration input σ_{RSM} increases from 7 to 18 G. This

(a)   (b)

(c)

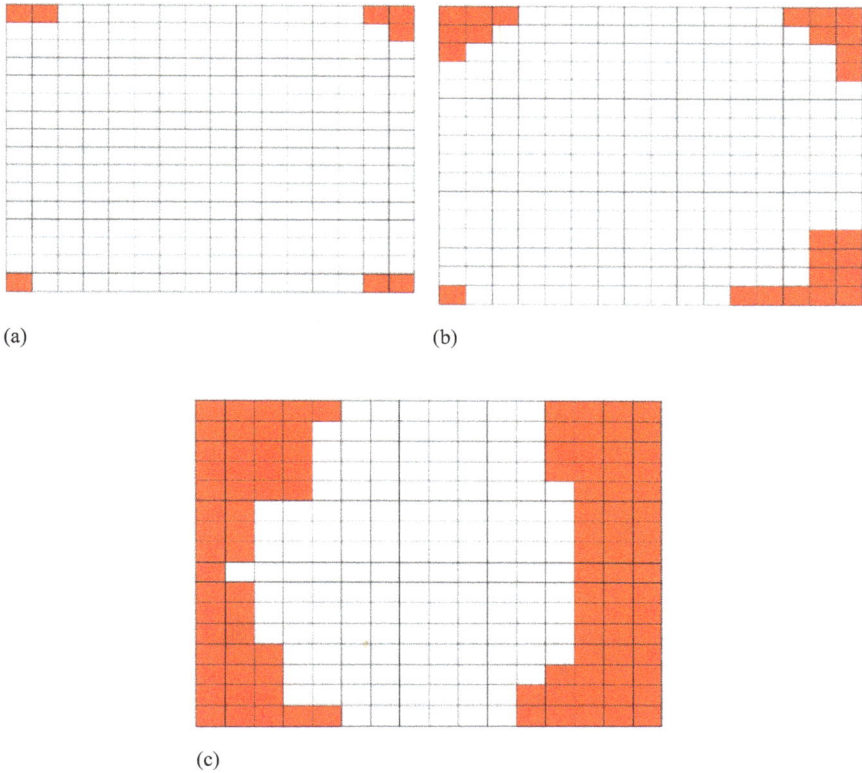

Figure 5.33   Failure solder joints distribution under the load condition of
G_{RMS}=18 G: (a) Failure solder joints distribution of component U1,
(b) failure solder joints distribution of component U2, and (c) failure
solder joints distribution of component U3

demonstrates that with increasing random vibration input levels, PBGA package solder joints are more likely to fail.

Figures 5.33 and 5.34 show the locations of solder joint failures from the random vibration experiments. The statistical results indicate that under the same input conditions, the number of failed solder joints increases as the position of the PBGA package chip moves from near the PCB edge to near the PCB center. This suggests that solder joints closer to the PCB center are more likely to fail than those closer to the PCB edge, consistent with the conclusions from the simulation analysis. In all the failed chips analyzed, a common factor is that the failed solder joints are located in the corner regions of the solder joint array, with no failures in the central region of the package, which is consistent with the simulation results. Furthermore, comparing Figures 5.33 and 5.34 shows that the number of failed solder joints increases with the level of random vibration input, which also aligns with the simulation analysis results.

(a)   (b)

(c)

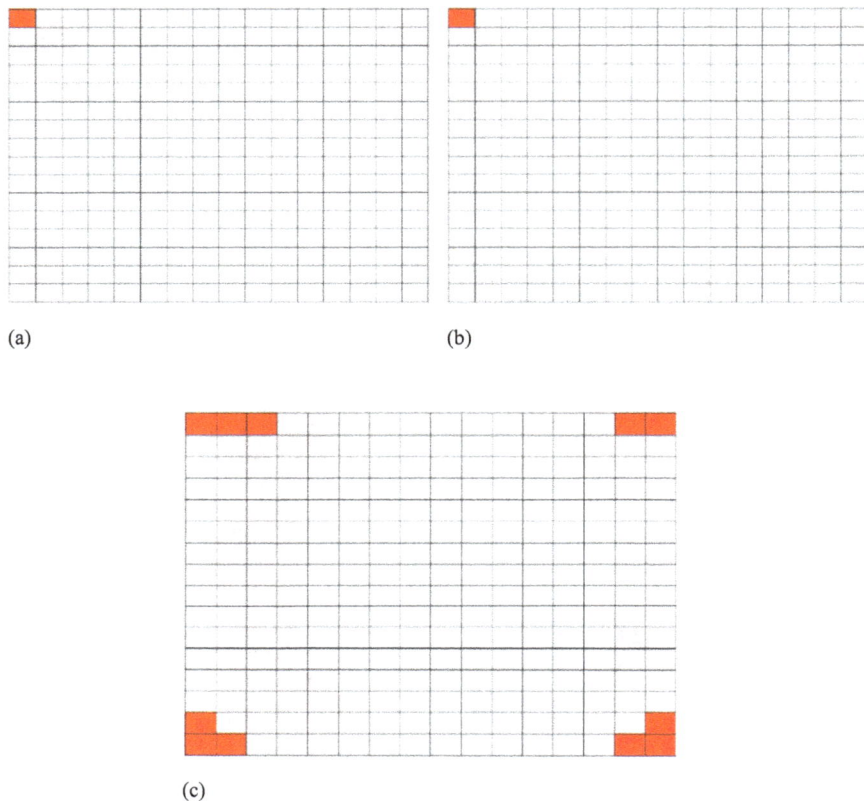

Figure 5.34   Failure solder joints distribution under the load condition of
            G_{RMS}=11 G: (a) failure solder joints distribution of component U12,
            (b) failure solder joints distribution of component U11, and
            (c) failure solder joints distribution of component U10

Through the stress analysis of the PBGA package solder joint array, it has been determined that under random vibration loads, the solder joints at the corner positions experience the highest stress and are most prone to failure. Figures 5.35, 5.36, and 5.37 show the stress distribution maps of key solder joints for various G_{RMS} loads of 7, 11, and 18 G, respectively. From these stress distribution maps, it can be observed that the maximum stress in these key solder joints is located on the side of the solder joint closest to the device and along the edge of the solder joint. This is similar to the results obtained from harmonic sweep vibration. Figure 5.38 presents the cross-sectional stress distribution map of the solder joint (1, A) under a G_{RMS} load of 11 G. The stress distribution map indicates that higher stress occurs along the edge of the solder joint closest to the device. During vibration, the alternating stress makes this location prone to fatigue cracks. As vibration continues, cracks will propagate along the weak interface of the solder joint until failure occurs. In

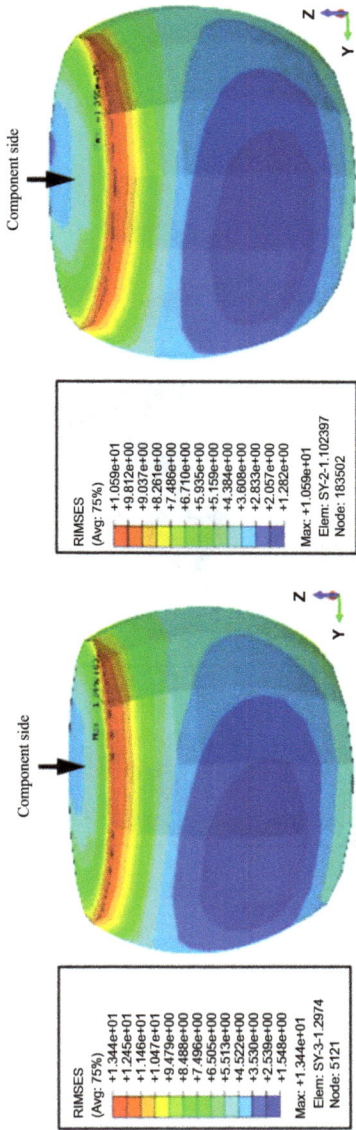

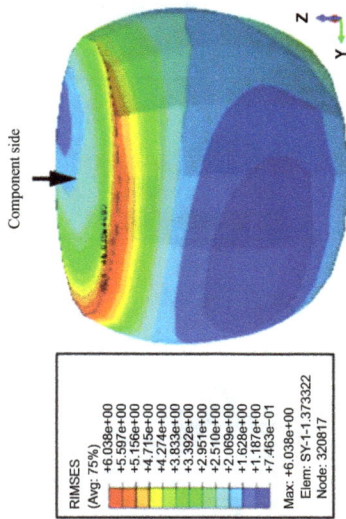

Figure 5.35   Stress contour of critical solder joint under the load condition of G_{RMS} = 7 G: (a) group 1, (b) group 2, and (c) group 3

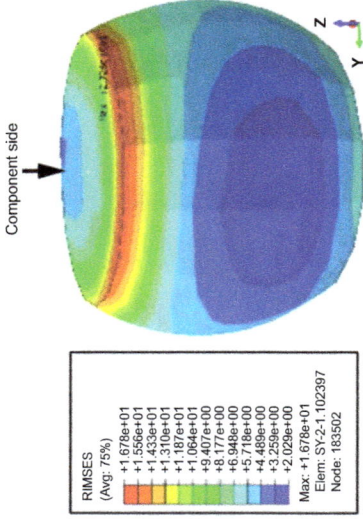

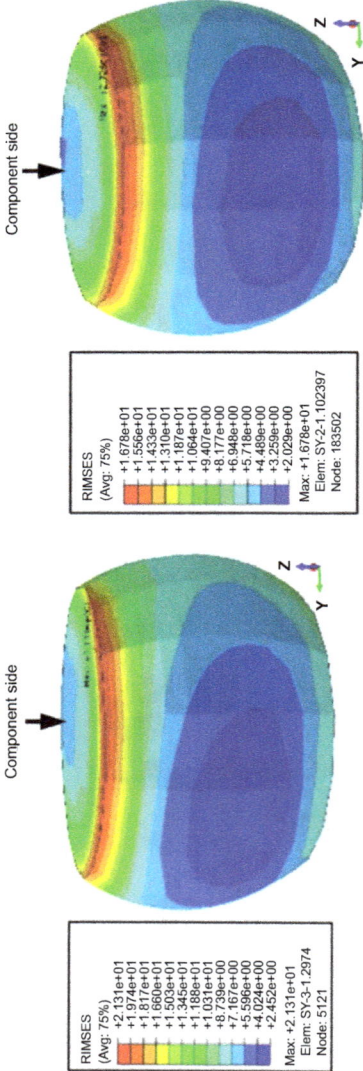

(a)

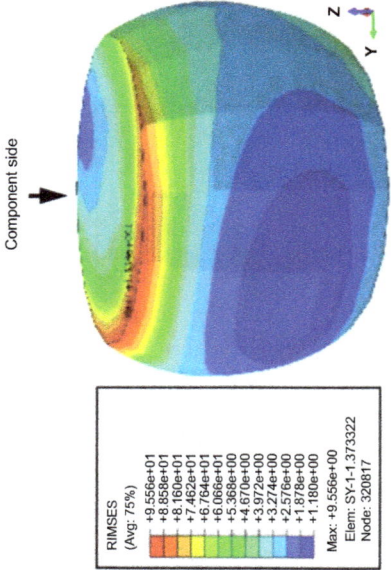

(b)

(c)

Figure 5.36   Stress contour of critical solder joint under the load condition of G_{RMS}=11 G: (a) group 1, (b) group 2, and (c) group 3

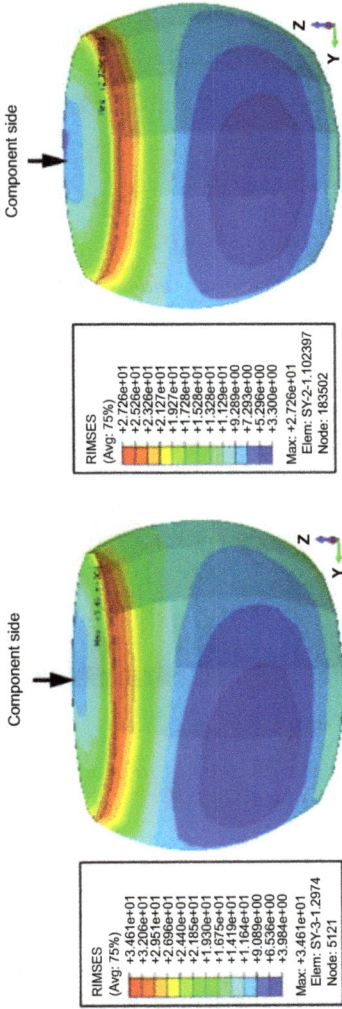

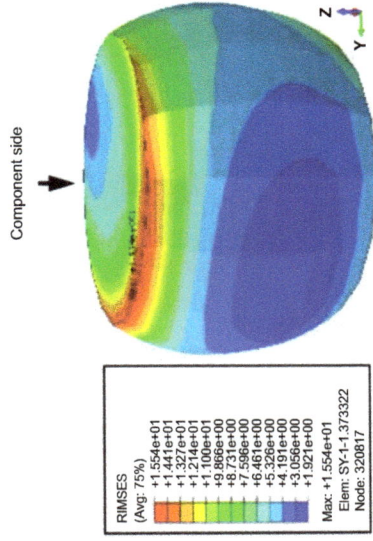

Figure 5.37 Stress contour of critical solder joint under the load condition of G_{RMS}=18 G: (a) group 1, (b) group 2, and (c) group 3

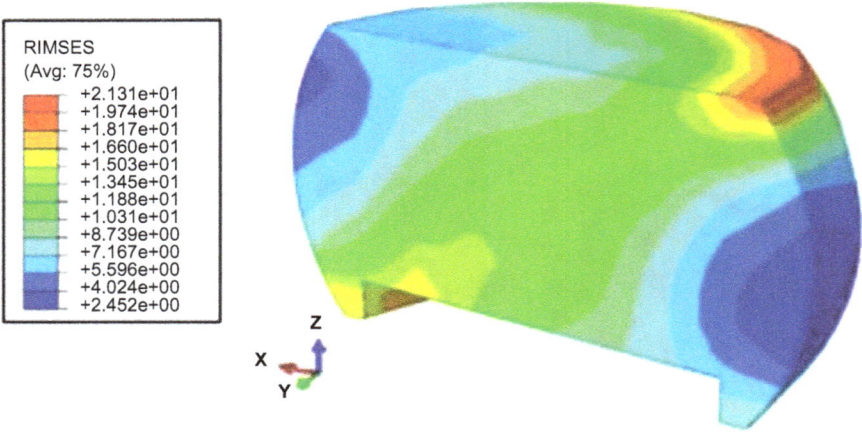

Figure 5.38   Cross-section of critical solder joint stress contour under random vibration

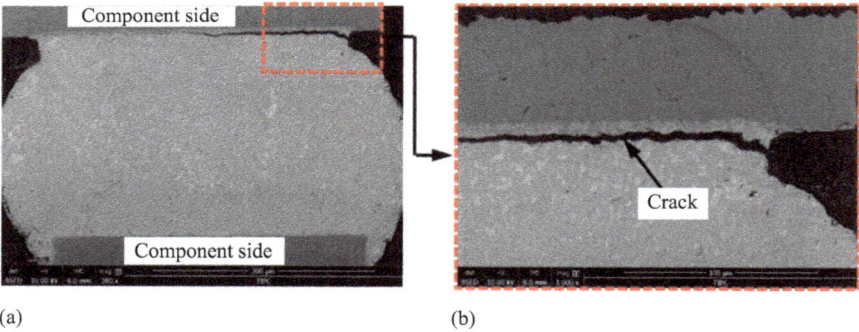

Figure 5.39   SEM image of failure solder joint: (a) morphology of failure solder joint (300×) and (b) crack morphology (1000×)

the random vibration test results, representative solder joint shapes and crack conditions are shown in Figure 5.39. The failed solder joint, located at a corner position of the solder joint array, exhibited a through-crack near the device side under a G_{RMS} load of 11 G. The crack extended from the edge of the solder joint toward the center, ultimately leading to joint fracture. The failure location and crack appearance in the solder joint are consistent with the numerical analysis results previously discussed.

In summary, under both types of vibration loads, the maximum stress in PBGA package solder joints increases as the chip position moves from the periphery toward the center of the PCB. The stress in solder joints in the central region of the PBGA package is lower, whereas the stress in the peripheral region is higher.

Solder joints at the corner regions exhibit significantly higher stress compared to other areas, and critical solder joints are all located at the corners. The maximum stress in these critical solder joints is consistently found on the side closest to the device. Based on these findings, it can be concluded that the stress response of PBGA under random vibration loads is similar to that under harmonic sweep vibration. During harmonic sweep vibration, the stress response in the solder joints is primarily influenced by first-order resonance when the sweep frequency is around the first natural frequency of the sample. Therefore, it is inferred that damage to the solder joints during random vibration, particularly at first-order resonance, is a major factor leading to solder joint failure, a conclusion also supported by the literature [8].

5.6 Chapter summary

This chapter presents an in-depth investigation into the stress behavior of PBGA package solder joints when subjected to vibrational loads. A finite element method (FEM) is used to simulate the stress distribution in the solder joints under two different types of vibrational loading: harmonic sweep and random vibration. The study provides a comprehensive numerical analysis that explores the impact of these vibrational forces on the structural integrity of the solder joints, which are critical for the reliability of PBGA packages in various practical applications, such as automotive, consumer electronics, and telecommunications.

The numerical simulations reveal several important findings. First, the results indicate that the maximum stress in the solder joints increases as the chip's position moves closer to the center of the PCB. Specifically, when the chip is positioned farther away from the PCB center, the solder joints experience lower stress levels. However, as the chip moves closer to the PCB center, the stress levels in the solder joints become significantly higher. This shift in stress distribution leads to an increased failure probability of the solder joints, particularly in the central regions. These observations are consistent with experimental results, which have shown that solder joints near the PCB center tend to fail more frequently under vibrational loading due to the higher stress they experience.

Further analysis shows that under vibrational loading, the solder joints located in the central region of the PBGA package experience relatively lower stress levels compared to those in the peripheral regions. In contrast, the solder joints situated at the corner points of the package experience much higher stress. As a result, the failure probability of solder joints at these corner points is notably higher than that of solder joints in other regions. This finding aligns with experimental data, which indicates that corner-point solder joints are more prone to failure under vibrational stress. The study also highlights that the critical solder joints, which are most likely to fail, are located at the corner points of the solder joint array. Moreover, the maximum stress in these corner joints occurs near the device side, supporting the experimental findings that solder joints at the corner, particularly near the device side, are more susceptible to failure.

The study also examines the effects of random vibrational loading on the PBGA package. It is found that the first-order resonance frequency plays a crucial role in the damage to the solder joints. The first-order resonance, which corresponds to a specific vibrational frequency, can amplify the stress experienced by the solder joints if the external vibrations coincide with this resonant frequency. This resonance phenomenon can lead to greater stress concentrations and a higher likelihood of failure. As a result, understanding and avoiding these resonant frequencies is essential for enhancing the durability of the PBGA package under vibrational loading conditions.

To address these challenges and improve the performance of PBGA package solder joints under vibrational stress, an experimental design approach was employed to optimize the vibrational behavior of the package. Through numerical simulations, the study identifies an optimal combination of parameters that reduce the stress in the solder joints and improve their reliability. The optimal design parameters found through this approach include a PCB thickness of 3 mm, a solder joint height of 0.40 mm, and a solder joint diameter of 0.66 mm. With these optimized parameters, the maximum stress in the solder joints is reduced to 8.12 MPa, representing a significant 40.29% reduction from the initial design (group 1), where the maximum stress was 13.60 MPa. This reduction in stress demonstrates the effectiveness of the design optimization in enhancing the vibrational performance of the PBGA package.

In conclusion, the findings from this study provide valuable insights into the stress distribution and failure mechanisms of PBGA package solder joints under vibrational loads. By identifying critical areas of stress concentration and understanding the influence of resonant frequencies, the study offers practical recommendations for optimizing the design of PBGA packages to enhance their durability and reliability. The successful optimization of key parameters, such as PCB thickness, solder joint height, and diameter, can significantly reduce the stress on the solder joints, improving the overall performance of the package. This research lays the foundation for further advancements in electronic packaging, particularly in applications where vibrational forces are a key factor in package reliability.

References

[1]   J.-E. Luan, T.Y. Tee, E. Pek, C.T. Lim, Z. Zhong, and J. Zhou, Advanced numerical and experimental techniques for analysis of dynamic responses and solder joint reliability during drop Impact, *IEEE Transactions on Components and Packaging Technologies* 29 (2006) 449–456.

[2]   T.L. Schmitz and K.S. Smith, *Mechanical vibrations: modeling and measurement*, Springer International Publishing, Cham, 2021.

[3]   M.C. Luintel, *Textbook of mechanical vibrations*, Springer Nature, Singapore, 2024.

[4] H. Benaroya and M. Nagurka, *Mechanical vibration*, CRC Press, Boca Raton, FL, 2009.

[5] Y.K. Kim, S. Lee, D. Hwang, and S. Jang, Analyses on the large size PBGA packaging reliability under random vibrations for space applications, *Microelectronics Reliability* 109 (2020) 113654.

[6] Y.K. Kim, S.M. Lee, D. Hwang, and S. Kim, High frequency effects on the PBGA stress developments at random vibration, in: *2019 22nd European Microelectronics and Packaging Conference & Exhibition (EMPC)*, IEEE, Pisa, Italy, 2019: pp. 1–4.

[7] Y.K. Kim and D.S. Hwang, PBGA packaging reliability assessments under random vibrations for space applications, *Microelectronics Reliability* 55 (2015) 172–179.

[8] W. Wang, Analysis on SMT lead-free solder joints reliability under random vibration loading, Master thesis, Shanghai Jiao Tong University, 2010.

Chapter 6

Electromigration simulation study of copper interconnects in 3D packaging

6.1 Introduction

As the dimensions of three-dimensional (3D) package micro-interconnects continue to shrink to the sub-micron scale, the associated reliability challenges become more pronounced. Among the key issues are the increased current density, elevated mechanical stress, and reduced heat dissipation, all of which can accelerate failure mechanisms in interconnects. Atomic-scale migration, a critical reliability phenomenon in ultra-large-scale integrated circuits (ULSI), is increasingly becoming a limiting factor in the performance and longevity of modern semiconductor devices. In this context, electromigration, where metal atoms move under the influence of an electric current, is one of the most significant contributors to the failure of interconnects. Copper (Cu), due to its low resistivity and superior resistance to electromigration compared to aluminum, has become the material of choice for next-generation interconnects in integrated circuits (ICs).

However, despite the advantages of Cu, the understanding of atomic migration and its impact on the reliability of Cu interconnects remains insufficient. Existing models for electromigration reliability analysis primarily focus on the study of single metal lines at a constant temperature, often under simplified assumptions. These models, while computationally efficient, are inadequate for capturing the complexities of real-world ICs, where temperature gradients, non-uniform current densities, and 3D interconnect structures play a crucial role in the electromigration process. In particular, the 3D nature of advanced ICs introduces significant variation in the distribution of current and temperature within interconnects, which influences the movement of metal atoms and, consequently, the risk of failure.

To address these limitations, this chapter introduces a more comprehensive approach to electromigration modeling by using COMSOL Multiphysics software. This method simulates the physical behavior of Cu interconnects in 3D, incorporating factors such as temperature gradients, varying current densities, and mechanical stresses. The finite-element method (FEM) is employed to accurately calculate the temperature, current density, and stress distributions within the interconnect structure. The results of these simulations offer deeper insights into the electromigration phenomena under realistic operational conditions, providing more accurate predictions of potential failure sites. By considering the full range of influencing factors, this approach provides a valuable tool for optimizing the design and reliability of Cu interconnects, which is critical for enhancing the overall performance and longevity of modern semiconductor devices. Ultimately, this new modeling method is expected to contribute to the development of more robust interconnect materials and design strategies, ensuring the continued scaling of ICs in the face of increasingly complex reliability challenges.

6.2   Theoretical model and simulation method of electromigration

The driving mechanisms of electromigration failure can be categorized into three main types: (1) migration induced by electrical currents (electronic migration), (2) migration driven by temperature gradients (thermal migration), and (3) migration caused by stress gradients (stress migration). Currently, the flux concept is commonly used to describe the driving effects of these three mechanisms in the process of electromigration. Electromigration failure is not an isolated phenomenon; it often occurs alongside thermal migration, stress migration, or chemical migration processes. Joule heating generated by high current densities creates temperature gradients, which serve as the driving force for thermal migration. Voids formed by electromigration induce internal stress in the interconnect structure. This stress, combined with mechanical and thermal loads, results in a stress gradient that drives atomic migration. Additionally, migration leads to non-uniform atomic concentrations within the structure, creating atomic concentration gradients that cause chemical migration. Therefore, electromigration failure is essentially a result of the combined effects of multiple migration mechanisms.

6.2.1   Electromigration theoretical model

Electromigration is the phenomenon where metal atoms diffuse along the direction of electron flow under high current density, essentially representing a process of atomic concentration redistribution. Fick's first law relates diffusion flux to concentration gradient, stating that under steady-state diffusion conditions, the mass of the diffusing substance (commonly referred to as diffusion flux) passing through a unit area perpendicular to the direction of diffusion per unit time is directly proportional to the concentration gradient at that section [1,2]

$$J_c = -D\nabla c \tag{6.1}$$

where J_c represents the diffusion flux caused by the concentration gradient (or chemical potential), c denotes the concentration, and ∇_c signifies the concentration gradient. The "−" sign indicates that the diffusion direction is opposite to the concentration gradient, meaning diffusion occurs from high- to low-concentration regions. $D = D_0 \exp(-E_A/k_B T)$ is the effective diffusion rate, where D_0 is the initial diffusion coefficient, E_A is the activation energy, k_B is the Boltzmann constant, and T is the absolute temperature.

In practice, most diffusion processes are non-steady-state, where the concentration of the diffusing substance changes over time. To study this situation, based on the mass balance of the diffusing substance, Fick's second law can be derived from Fick's first law

$$\frac{\partial c}{\partial t} = \nabla[-D\nabla c] \tag{6.2}$$

Under thermodynamic equilibrium conditions, the atomic concentration follows the Boltzmann distribution

$$c(x) = c_0 \exp\left(\frac{-\varphi(x)}{k_B T}\right)$$

where $\varphi(x)$ is the potential energy, and the derivative of $c(x)$ yields

$$\frac{dc}{dx} = -c_0 \exp\left(\frac{-\varphi(x)}{k_B T}\right) \frac{1}{k_B T} \frac{d\varphi}{dx} = -\frac{c}{k_B T} \frac{d\varphi}{dx}$$

where $-\frac{d\varphi}{dx}$ is the potential energy difference or potential energy gradient caused by the concentration gradient. This gradient can create a driving force that "pushes" the substance to migrate from high potential energy to low potential energy, denoted as F_c.

Therefore, Fick's first law can also be expressed as

$$J = D\frac{c}{k_B T} F_c \tag{6.3}$$

"Electron wind force" refers to the force exerted on metal atoms due to momentum exchange from collisions with electrons under the influence of an electric field. Its direction is the same as that of the electron motion. Coulomb force refers to the direct force exerted by the electric field on metal atoms, and its direction is parallel to the electric field. Generally, the combined force of the electron wind force and Coulomb force is collectively referred to as the "electron wind force." The atomic flux induced by "electron wind force" can be expressed as

$$J_{EM} = \frac{cD}{k_B T} Z^* e\rho j \tag{6.4}$$

where c represents the atomic concentration, e is the electron charge, Z^* is the effective charge, j is the current density vector, and $\rho = \rho_0[1 + \alpha(T - T_0)]$, which is the temperature-dependent resistivity, where α is the temperature coefficient of the metal material and ρ_0 is the resistivity corresponding to the initial temperature T_0.

From this formula, it can be seen that the current density is one of the main factors influencing electromigration. The higher the current density, the more atoms undergo electromigration, leading to a faster electromigration failure process.

Thermal migration has become a potential hazard leading to solder joint failure. Thermal migration refers to the directional movement of metal atoms under the influence of a temperature gradient, ultimately causing phase separation of interconnect material metals and damaging the microstructure. Atoms tend to move from higher temperature regions to lower temperature regions, so a temperature gradient can induce atomic migration and create a driving force. Joule heating is one of the factors most likely to create a temperature gradient on the interconnect surface. This temperature gradient already exists before electromigration occurs, and during the formation of voids and hillocks, localized Joule heating effects accelerate the increase in the temperature gradient associated with electromigration [3–5]. The atomic flux induced by the temperature gradient can be expressed as

$$J_{\text{TM}} = -\frac{cD}{k_B T} \frac{\nabla T}{T} Q^*$$

(6.5)

where Q^* represents the molar heat flux, which can be defined as the difference in energy (enthalpy) per mole of atoms in motion compared to their initial state. Equation (6.2) indicates that the greater the temperature gradient, the larger the atomic flux.

Stress migration refers to the phenomenon of directional movement of metal atoms under a certain stress gradient, ultimately leading to the formation of microvoids and cracks. The sources of the stress gradient mainly come from two aspects: first, the "backflow stress" induced by electromigration, which exists throughout the entire process of electromigration; second, the non-uniform stress distribution caused by the mismatch in thermal expansion coefficients between the metal interconnect materials and the surrounding materials [6]. This stress gradient occurs at the onset of electromigration and will be released once voids are formed. The atomic flux induced by the stress gradient can be expressed as

$$J_{SM} = \frac{cD}{k_B T} \Omega \nabla \sigma_H$$

(6.6)

where Ω represents the atomic volume and σ_H is the hydrostatic pressure.

$$\sigma_H = \frac{\sigma_{xx} + \sigma_{yy} + \sigma_{zz}}{3}$$

(6.7)

where σ_{xx}, σ_{yy}, σ_{zz} represent the stress components in each principal direction.

The migration direction of metal atoms is always from the region with the highest stress gradient. In areas with concentrated stress, the stress gradient is significant. When the driving force generated by this stress gradient exceeds the force required for atomic migration, migration will occur in the stress-concentrated areas, leading to the formation of microvoids and cracks, ultimately resulting in interconnect failure.

6.2.2   Electromigration simulation method based on atomic flux divergence

Electro-thermal coupling simulations can only provide distributions of current density and temperature gradient within solder joints. Judging the electromigration failure location based on the maximum current density is not accurate, as various migration mechanisms affect the atomic flux in interconnect structures. This section introduces the atomic flux divergence method to describe the electromigration process through the divergence of the three main driving forces: electron wind force, temperature gradient, and stress gradient. It also calculates the atomic flux divergence distribution in the solder joint and the electromigration failure time. The location of the maximum atomic flux divergence represents the weakest part of the metal interconnect structure, where void formation and failure are likely to occur first.

Considering electromigration driven by electron wind force, temperature gradient, and stress gradient, while neglecting the effect of atomic concentration gradients, the total flux of migrating atoms under multiple field couplings is the sum of the atomic fluxes due to each driving force. Therefore, the expression for the total flux of migrating atoms can be written as

$$J = J_{EM} + J_{TM} + J_{SM} = \frac{cD}{k_B T} Z^* e\rho j - \frac{cD}{k_B T} \frac{\nabla T}{T} Q^* + \frac{cD}{k_B T} \Omega \nabla \sigma_H \qquad (6.8)$$

Electromigration is a diffusion-controlled mass transport process. The time-dependent evolution equation for atomic concentration caused by an applied current is a classical mass conservation equation:

$$\nabla \cdot J + \frac{\partial c}{\partial t} = 0 \qquad (6.9)$$

where J represents the total flux of migrating atoms and t denotes time.

Assuming that the current density j is constant, neglecting the divergence of the current density, the divergence of the atomic flux caused by the "electron wind force" is given by

$$\text{div}(J_{EM}) = \left(\frac{E_A}{k_B T} - \frac{1}{T} + \frac{\rho_0}{\rho} \right) \cdot J_{EM} \cdot \nabla T + \frac{J_{EM}}{c} \cdot \nabla c \qquad (6.10)$$

The atomic flux divergence caused by the temperature gradient is given by

$$\text{div}(J_{TM}) = \left(\frac{E_a}{k_B T} - \frac{2}{T} \right) \cdot J_{TM} \cdot \nabla T - \frac{cD}{k_B T} Q^* \text{div}(\nabla T) + \frac{J_{TM}}{c} \cdot \nabla c \qquad (6.11)$$

The atomic flux divergence caused by the stress gradient is given by

$$\text{div}(J_{SM}) = \left(\frac{E_a}{k_B T} - \frac{2}{T} \right) \cdot J_{SM} \cdot \nabla T - \frac{cD}{k_B T} Q^* \text{div}(\nabla \sigma_H) + \frac{J_{SM}}{c} \cdot \nabla c \qquad (6.12)$$

In summary, neglecting the divergence of atomic concentration, the total atomic flux divergence caused by the electron wind force, temperature gradient, and stress gradient is given by

$$
\begin{aligned}
\text{div}(J) &= \text{div}(J_{EM}) + \text{div}(J_{TM}) + \text{div}(J_{SM}) \\
&= \left(\frac{E_a}{k_B T} - \frac{1}{T} + \frac{\rho_0}{\rho}\right) \cdot J_{EM} \cdot \nabla T + \left(\frac{E_a}{k_B T} - \frac{2}{T}\right) \cdot J_{TM} \cdot \nabla T \\
&\quad - \frac{cD}{k_B T} Q^* \text{div}(\nabla \sigma_H) + \left(\frac{E_a}{k_B T} - \frac{2}{T}\right) \cdot J_{SM} \cdot \nabla T + \frac{cD}{k_B T} Q^* \text{div}(\nabla \sigma_H)
\end{aligned}
$$

$$
(6.13)
$$

The above equation can be rewritten as

$$
\text{div}(J) = c \cdot H(T, \sigma, j, \ldots) \tag{6.14}
$$

From (6.7), it can be seen that the atomic flux divergence is directly proportional to the atomic concentration c and H is a function dependent on physical parameters such as temperature, stress, electric field, and atomic concentration.

$$
c_{i+1} = c_i e^{-H_i \Delta t_i} \tag{6.15}
$$

where i represents the discrete nodes, with i = 1, 2, ..., n. c_i denotes the atomic concentration at the discrete node i, and H_i represents the value of the function H at the discrete node i.

6.3   Case study and analysis

6.3.1   Establishment of the FEM

In this section, a geometric model including leads is established using COMSOL Multiphysics software. This includes the meshing method, selection of physical fields and their parameters, as well as the results of coupling various physical fields. This section will simulate the current density, temperature, and stress of the Cu interconnect lines under different voltage inputs, comparing the effects of varying input voltages on the electromigration, thermo-migration, and stress migration of the Cu interconnect lines. The tree structure of the interconnect lines studied is 3D; therefore, a 3D solid structure is used for modeling. The interconnect structure established in this work is shown in Figure 6.1, with the parameters of the interconnect and substrate structure presented in Table 6.1.

The interconnect material is set to Cu from the software's built-in material library, and the substrate material is silicon (Si) (single-crystal, isotropic) from the micro-electromechanical system (MEMS) module. The material parameters are shown in Table 6.2.

The meshing is performed using COMSOL's automatic free tetrahedral mesh method. The maximum element size is 4.4 nm, the minimum element size is 0.32 mm, the maximum element growth rate is 1.4, the curvature factor is 0.4, and the resolution for narrow regions is 1.7. The meshing is shown in Figure 6.2.

Based on the previous analysis, it is evident that the analysis of the interconnect tree structure involves electric field analysis, stress analysis, and heat transfer analysis. Therefore, this work uses an electro-thermal coupled multi-physics model. The

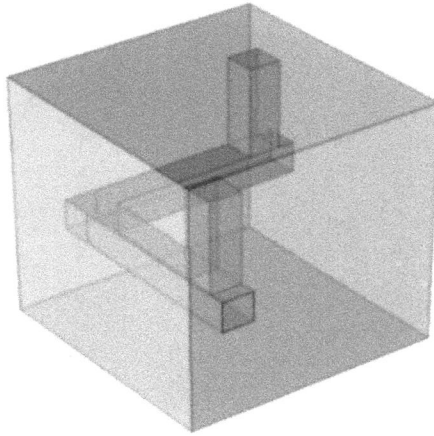

Figure 6.1 3D geometric model

Table 6.1 Parameters of line structure

Model	Length (nm)	Width (nm)	Thickness (nm)
Vertical interconnect line	10	10	30
Horizontal interconnection line	70	10	10
Substrate filling	80	80	80

Table 6.2 Material parameters

Materials	Cu	Si
Electrical conductivity (S/m)	5.998×10^7	–
Thermal expansion coefficient (1/K)	17×10^{-6}	2.6×10^{-6}
Constant-pressure heat capacity (J/(kg K))	5	11.7
Relative dielectric constant	8960	2329
Thermal conductivity (W/(m K))	400	130
Young's modulus (Pa)	110×10^9	170×10^9
Poisson's ratio	0.35	0.28

electric field module calculates the electric field, current, and potential distribution in the conductor dielectric. The heat transfer module determines the temperature distribution and temperature gradient in the material, while the solid mechanics module calculates the stress distribution in the model. The key parameters obtained are used in (6.12) to characterize the degree of electromigration.

The initial temperature is set to room temperature, 293.15 K. This work applies a constant temperature to the bottom surface, while the top surface is subjected to convective heat transfer to simulate the heat dissipation on the circuit board's

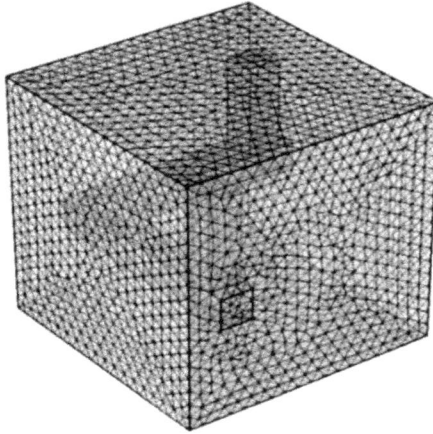

Figure 6.2   Meshing of geometric structure

surface. Vertical surfaces use thermal insulation boundary conditions to represent the heat dissipation at a specific location on the circuit board. The convective heat flux equations are given by (6.16) and (6.17), where C_p is the heat capacity at constant pressure, k is the thermal conductivity, and h is the heat transfer coefficient, with a value of 5 W/(m^2·K), and T_{ext} is the external temperature, 293.15 K

$$\rho C_{\mathrm{p}} \frac{\partial T}{\partial t} + \rho C_{\mathrm{p}} u \cdot \nabla T + \nabla \cdot q = Q + Q_{ted} \tag{6.16}$$

$$q = h \cdot (T_{ext} - T) \tag{6.17}$$

In this work, the current is conducted through the Cu interconnect lines, with the potential applied to the top surface of the Cu wires (0.25 and 0.3 V) to compare the effects of input potential on the distribution of physical fields and atomic flux divergence. The bottom surface is grounded. The equations for the AC/DC module are given by (6.18) and (6.19), where ε denotes the relative permittivity

$$\nabla \cdot J = Q_j J = \left(\sigma + \varepsilon_0 \varepsilon_0 \frac{\partial}{\partial T} \right) E + J_{\mathrm{e}} \tag{6.18}$$

$$E = -\nabla V \tag{6.19}$$

The calculation is performed by applying fixed boundary conditions on the bottom edge. Both the wire and the substrate material are considered linear elastic materials. The calculation uses a transient study with a total duration of 1 h and a time step of 0.1 h.

6.3.2   Results and discussion

Figure 6.3 shows the temperature distribution in the Cu interconnect lines under different input voltages. Figure 6.4 presents the isothermal surface plots for the Cu

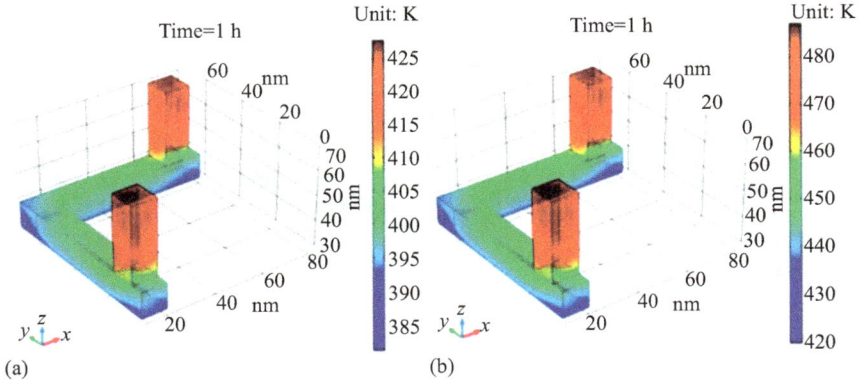

Figure 6.3   Temperature distribution in Cu interconnect lines: (a) 0.25 V and (b) 0.3 V

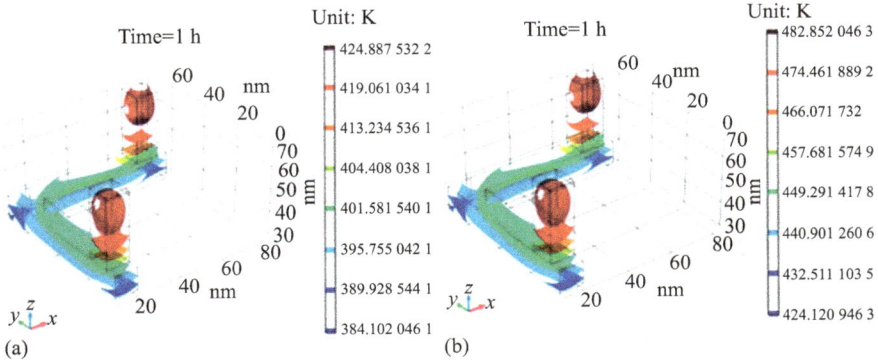

Figure 6.4   Intermediate temperature distribution of Cu interconnect lines: (a) 0.25 V and (b) 0.3 V

interconnect lines at two input voltages. From Figure 6.3, it can be observed that the temperature of the interconnect lines below the Cu interconnect line is the lowest. When the input voltage is 0.25 V, the minimum temperature is 384.1 K; when the input voltage is 0.3 V, the minimum temperature is 424 K. As the distance from the top surface decreases, the temperature gradually increases, with the maximum temperature occurring at the current input port. When the input voltage is 0.25 V, the highest temperature is 424.9 K. When the input voltage is 0.3 V, the highest temperature is 482.5 K.

Figure 6.5 shows the volumetric distribution of current density in the Cu interconnect lines under different input voltages. The figure indicates that the current density remains nearly constant in the straight sections of the interconnect lines but experiences abrupt changes at the bends. The current density is highest at the corners of the Cu interconnect lines, reaching up to 1.8×10^{14} A/m^3 when the input voltage is

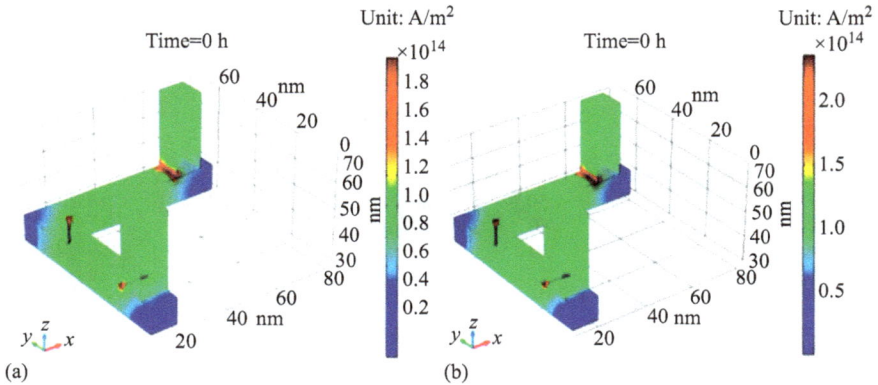

Figure 6.5   Current density distribution of Cu interconnect lines: (a) 0.25 V and (b) 0.3 V

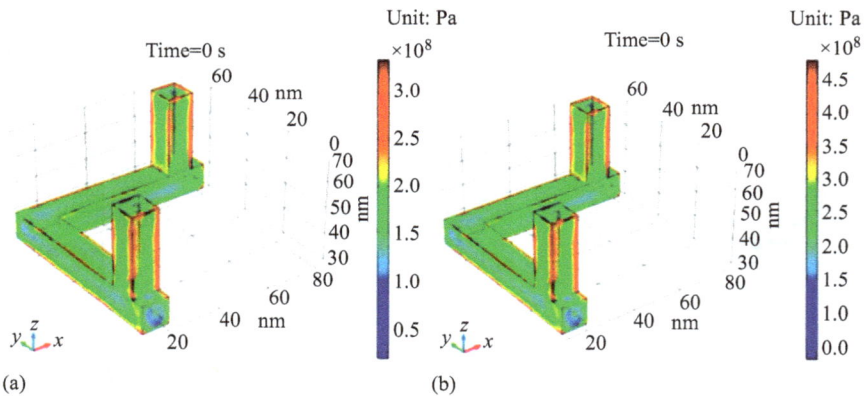

Figure 6.6   Stress distribution diagram of Cu interconnect lines

0.25 V. When the input voltage is increased by 20%, the current density also increases by 20%, and the calculation results are consistent with the actual observations.

Figure 6.6 shows the stress distribution in the Cu interconnect lines under different voltages. According to Table 6.2, the thermal expansion coefficient of Cu is 17×10^{-6} K^{-1}, while that of Si is 2.6×10^{-6} K^{-1}. The strong thermal mismatch between Cu and Si results in significant stress at the edges of the Cu. Under an input voltage of 0.25 V, the maximum stress reaches 3.72×10^8 N/m^2; under an input voltage of 0.3 V, the maximum stress reaches 4.76×10^8 N/m^2. When the input voltage is increased by 20%, the stress increases by 27.96% due to the coupling of temperature effects.

By performing a multi-physics coupled finite-element analysis of Cu interconnect lines using COMSOL Multiphysics, the temperature distribution, current distribution, and stress field distribution of the Cu interconnect lines under different input voltages were obtained. The results of the finite-element simulations are incorporated into the atomic diffusion flux divergence equation (6.16) to compare and analyze electromigration, thermos-migration, and stress migration in Cu interconnect lines under different input voltages. Table 6.3 lists the parameters required for the Cu interconnect line calculations.

Figure 6.7 illustrates the atomic diffusion flux divergence distribution caused by electromigration under different input voltages. The results indicate that the atomic diffusion flux divergence caused by electromigration is nearly zero in the straight sections of the interconnect lines but is significant at the bends. At the current inflow direction of the bend, the atomic flux divergence is positive, indicating atomic migration; whereas, at the current outflow direction of the bend, the atomic flux divergence is negative, indicating atomic deposition, which aligns with previous observations. When the input voltage is increased by 20%, the current density in the Cu interconnect lines increases by 20% as well, while the atomic diffusion flux divergence increases by a factor of 12.6. The results demonstrate that

Table 6.3   Cu interconnect wire electromigration parameters

Name	Symbol	Value	Unit
Boltzmann constant	k_B	1.38×10^{-30}	J/K
Heat transfer coefficient	Q	377	$W/(m^2 \; K)$
Lattice constant	Ω	8.78×10^{-30}	m^3
Effective charge number	Z	2.5	1
Resistivity	ρ	1.67×10^{-8}	$\Omega \; m$
Pre-exponential factor	D_0	7.56×10^{-5}	m^2/s
Electron charge	e	1.6×10^{-19}	C
Avogadro's constant	N	6.02×10^{23}	1

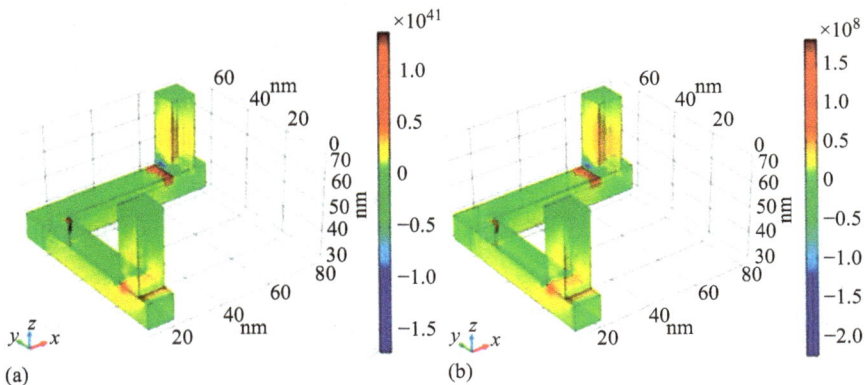

(a)

(b)

Figure 6.7   Atomic diffusion flux dispersion due to electron migration in Cu interconnect lines: (a) 0.25 V and (b) 0.3 V

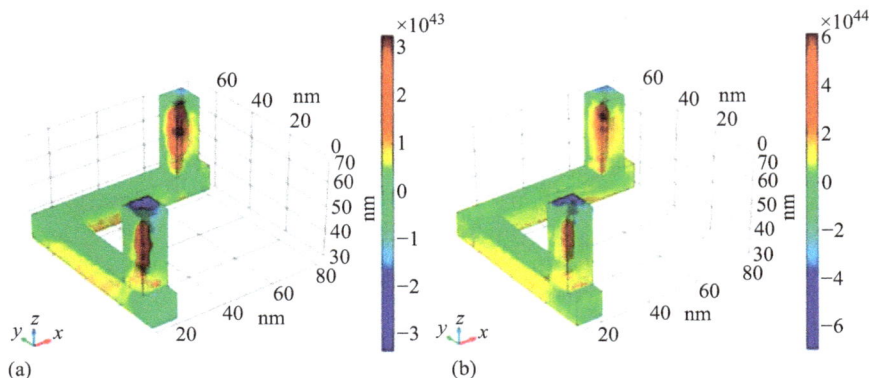

Figure 6.8   *Distribution of thermal migration atomic diffusion flux dispersion of Cu interconnects: (a) 0.25 V and (b) 0.3 V*

due to the coupling of multiple physical fields, the input voltage has a substantial impact on the atomic diffusion flux divergence caused by electromigration. Therefore, carefully selecting the power supply parameters and heat dissipation conditions for ICs is crucial for ensuring the rationality of circuit design and effectively mitigating electromigration.

Figure 6.8 shows the distribution of thermal migration atomic diffusion flux divergence in Cu interconnect lines under different voltages. The figure indicates that the values of thermal migration atomic diffusion flux divergence are higher in regions with higher temperatures. In the Cu interconnect lines, thermal migration is more severe in the higher temperature regions and in the vertical wire regions. As shown in Figure 6.5, the temperature gradient is larger in the vertical wire regions, which results in a greater driving force for atomic diffusion in these areas. Comparing the cases of input voltages of 0.25 and 0.3 V, it is observed that an increase in voltage leads to a rise in wire temperature and an increase in the temperature gradient of the wire, making the thermal migration significantly more pronounced.

First, a comparison of the physical field distributions of different interconnect materials is conducted. Figure 6.9 shows the temperature distribution and iso-thermal surfaces for different materials. From Figures 6.9–6.11, it can be observed that under the same input voltage (0.25 V), the highest temperature in Cu inter-connect lines reaches 424 K, the maximum current density is 1.91×10^{14} A/m^2, and the maximum stress is 3.74×10^8 Pa. For Ag interconnect lines, the highest temperature is 430 K, the maximum current density is 1.63×10^{14} A/m^2, and the maximum stress is 3.11×10^8 Pa. The physical field distributions in Cu and Ag interconnect lines are similar. This is primarily because both Ag and Cu materials have high electrical and thermal conductivities, resulting in similar physical field distributions under the same input voltage. This study incorporates the calculated physical field distributions into the atomic diffusion flux divergence equation to simulate and compare the electromigration characteristics of Ag and Cu wires.

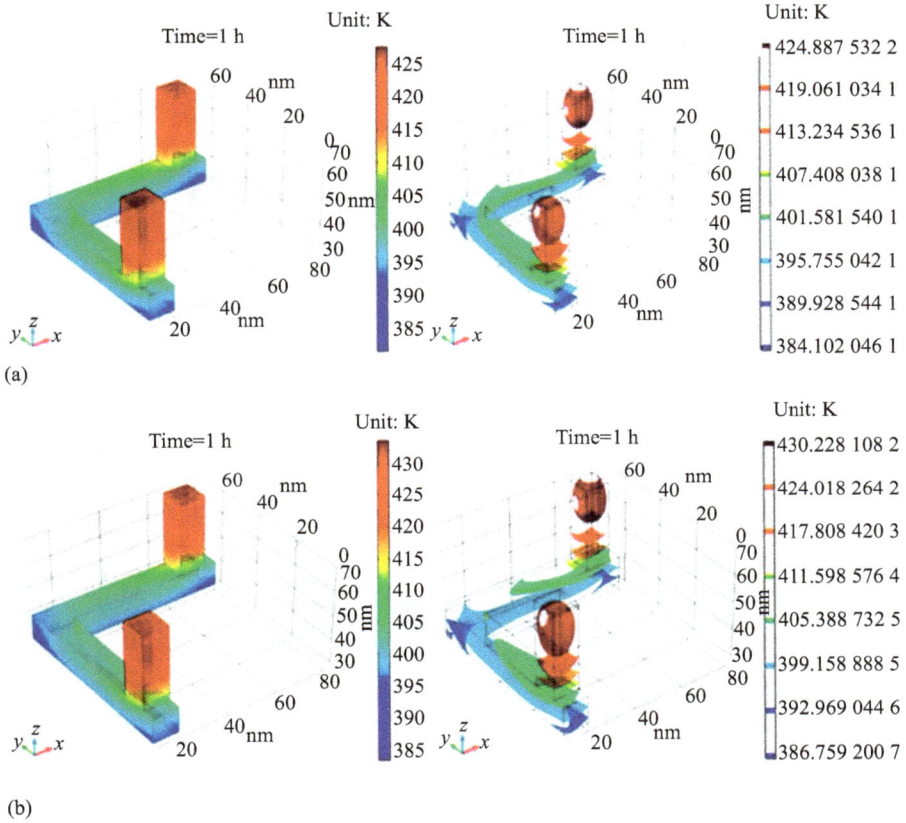

Figure 6.9   *Temperature distribution diagram and isothermal surface of (a) Cu interconnect line and (b) Ag interconnect line*

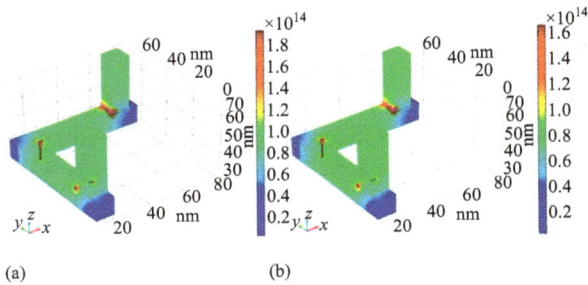

Figure 6.10   *Current density distribution of (a) Cu interconnect line and (b) Ag interconnect line*

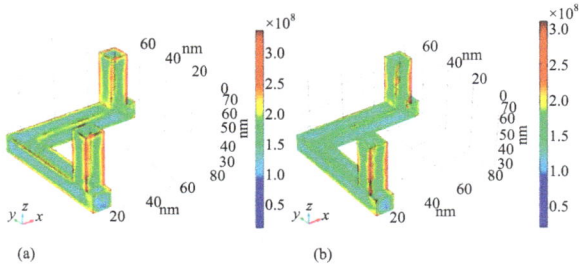

Figure 6.11   Stress distribution diagram of (a) Cu interconnect line and (b) Ag interconnect line

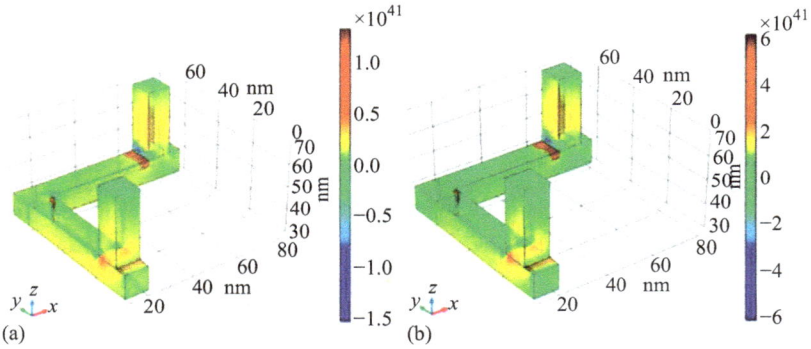

Figure 6.12   Comparison of electromigration of (a) Cu interconnect line and (b) Ag interconnect line

Figure 6.12 shows a comparison of the electromigration atomic diffusion flux divergence for Cu and Ag materials. As seen in Figure 6.12, at an input voltage of 0.25 V, electromigration is more severe in Ag interconnect lines compared to Cu interconnect lines. The main reason is that silver has a lower melting point and a higher number of effective valence charges, making it more prone to electromigration. Therefore, Cu interconnect lines have better resistance to electromigration than Ag interconnect lines.

Figure 6.13 presents a comparison of the thermal migration performance between Cu and Ag interconnect lines. According to the simulation results, the maximum atomic flux divergence for Cu interconnect lines is 3.19×10^{43}, while the thermal migration atomic diffusion flux divergence for Ag is 4.78×10^{43}. Under the same input voltage, the thermal migration at the maximum migration point for Cu interconnect lines is only superior to that of Ag interconnect lines.

Figure 6.14 presents a comparison of stress migration between Cu and Ag interconnect lines. The calculation results show that, at an input voltage of 0.25 V, the maximum stress migration atomic flux divergence is nearly the same for both

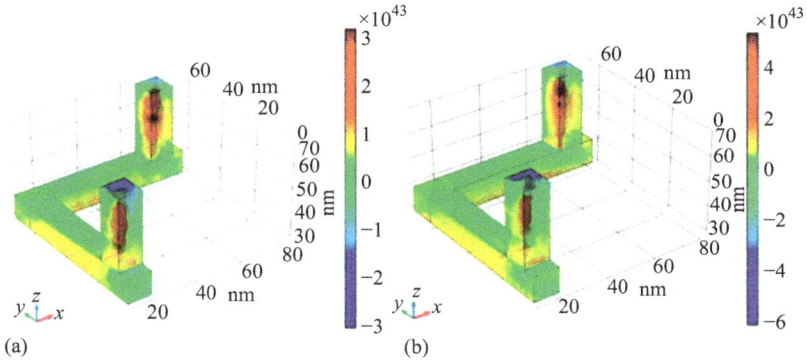

Figure 6.13   Comparison of thermal migration: (a) Cu and (b) Ag

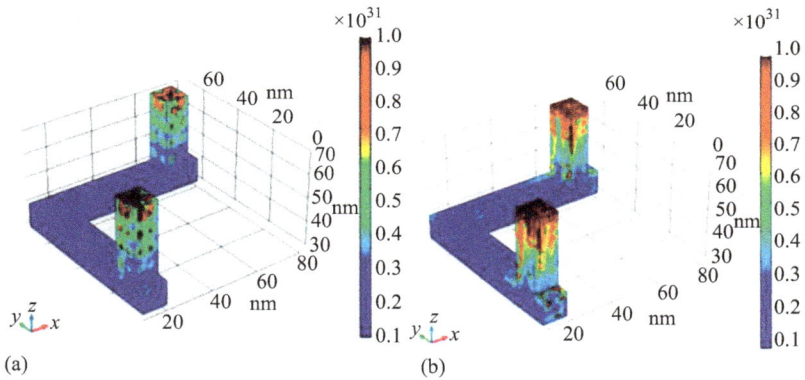

Figure 6.14   Comparison of stress migration of interconnection lines: (a) Cu and (b) Ag

Cu and Ag interconnect lines, with Cu at 1.0×10^{31} and Ag at 0.97×10^{31}. Therefore, Cu interconnect lines exhibit overall better electromigration resistance compared to Ag interconnect lines, making Cu a superior material for high-density integrated circuit conductors [7,8].

6.4   Chapter summary

The simulation results demonstrate a significant accumulation of current at the inner corners of the metal interconnect lines, a phenomenon that is central to understanding the behavior of electromigration in these structures. Electromigration is most severe at these corners due to the highly localized current density, which accelerates the movement of atoms in the metal. Specifically, atomic migration occurs in two distinct directions: atomic accumulation in the direction of current inflow and atomic depletion in the direction of current outflow. This directional flow of atoms contributes to the degradation of the interconnect line over time, leading to the formation of voids at the outflow side and material buildup at the inflow side, which can ultimately cause failure in the interconnect.

Another key observation from the simulations is that the high-temperature region is located between the inner and outer sides of the corner of the interconnect lines. Temperature distribution within these lines is primarily governed by the current density and the heat dissipation characteristics of the underlying circuit board. As current flows through the interconnects, resistive heating occurs, which leads to an increase in the local temperature. This increase in temperature significantly amplifies the risk of migration failure, as higher temperatures accelerate atomic diffusion, making it easier for atoms to migrate and create voids or hillocks. Therefore, a robust cooling system is essential to maintain the temperature of the interconnect lines within safe limits, thereby reducing the risk of electromigration-related failures. Proper thermal management can greatly enhance the overall reliability of metal interconnects by preventing excessive heating and ensuring efficient heat dissipation.

Furthermore, the simulations indicate that stress is concentrated at the outer edges of the corners of the interconnect lines. This localized stress is a result of the geometric design of the interconnects and the mechanical forces acting on the materials. However, the contribution of stress migration to overall failure is relatively minimal in this model. The impact of stress migration on the degradation of the interconnects is almost negligible, which suggests that while stress concentration at the corners is an important factor, it does not significantly influence the migration process in this case. The primary concern remains the effect of electromigration and thermal migration, which are driven by the current density and temperature gradients.

Another important finding is the comparison between different materials for the interconnect lines. Under the same operating conditions, Cu interconnects exhibit superior performance in terms of both electromigration resistance and thermal migration resistance compared to silver (Ag) interconnects. Cu's superior properties make it the preferred material for high-density electronic circuits, as it can withstand thermal and electrical stresses more effectively than silver. In particular, Cu's better electromigration resistance means it is less likely to experience atomic migration under high current densities, while its improved thermal migration resistance helps to mitigate the effects of localized heating. Therefore, Cu interconnects are more suitable for advanced electronic packaging processes where both electromigration and thermal migration are critical concerns.

In conclusion, the study emphasizes the importance of material choice, geometric design, and thermal management in improving the reliability of metal interconnects in electronic circuits. While stress concentration at the corners plays a role, the primary failure mechanism remains electromigration, which can be mitigated through careful design and material selection, such as the use of Cu interconnects. Additionally, efficient cooling strategies are vital in reducing the temperature and ensuring the long-term durability of interconnects in high-performance electronic devices.

References

[1] P. Liu, A. Overson, and D. Goyal, Electromigration mechanisms of solder joints with limited Sn volume in advanced electronic packaging, in: *2021 IEEE 71st Electronic Components and Technology Conference (ECTC)*, IEEE, San Diego, CA, USA, 2021: pp. 918–924.

[2] C.-L. Liang, Y.-S. Lin, C.-L. Kao, *et al.*, Electromigration reliability of advanced high-density Fan-Out packaging with fine-pitch 2-/2-μm L/S Cu redistribution lines, *IEEE Transactions on Components and Packaging Technologies* 10 (2020) 1438–1445.

[3] L. Chen, S.X.-D. Tan, Z. Sun, S. Peng, M. Tang, and J. Mao, A fast semi-analytic approach for combined electromigration and thermomigration analysis for general multisegment interconnects, *IEEE Transactions on Computer-Aided Design of Integrated Circuits and Systems* 40 (2021) 350–363.

[4] Y. Du, Y. Qiao, X. Ren, Y. Lai, and N. Zhao, Characterization of Sn-xIn solders and thermomigration-induced interfacial IMC growth of Cu/Sn-xIn/Cu micro solder joints, *Electronics* 12 (2023) 1899.

[5] C. Chen, H.-Y. Hsiao, Y.-W. Chang, F. Ouyang, and K.N. Tu, Thermomigration in solder joints, *Materials Science and Engineering: Reports* 73 (2012) 85–100.

[6] P.K. Huang, C.Y. Lu, W.H. Wei, *et al.*, Wafer level system integration of the fifth generation CoWoS®-S with high performance Si interposer at 2500 mm^2, in: *2021 IEEE 71st Electronic Components and Technology Conference (ECTC)*, IEEE, San Diego, CA, USA, 2021: pp. 101–104.

[7] S. Zhang, X. Qi, M. Yang, *et al.*, A study on the resistivity and mechanical properties of modified nano-Ag coated Cu particles in electrically conductive adhesives, *Journal of Materials Science: Materials in Electronics* 30 (2019) 9171–9183.

[8] Y.-S. Kim, S. Zhang, and K.-W. Paik, Highly reliable solder ACFs FOB (flex-on-board) interconnection using ultrasonic bonding, *Journal of the Microelectronics and Packaging Society* 22 (2015) 35–41.

Chapter 7

Study on the influence of mechanical properties of TSV-Cu on cracking in TSV/RDL interconnect structures

7.1 Introduction

The expansion of through-silicon vias copper (TSV-Cu) and the presence of internal residual stress in interconnect layers can significantly impact the mechanical integrity and reliability of electronic packaging. These stresses often lead to deformation, void formation, delamination, and cracking in interconnect layers, especially at the interfaces between different materials such as TSV-Cu and redistribution layer (RDL) [1,2]. Existing literature tends to focus on failure localization and observation during reliability testing of TSV/RDL interconnects, but often lacks detailed investigations into the underlying failure mechanisms, particularly with respect to how these interconnects respond under different loading conditions over time.

Given the high costs and time constraints associated with experimental testing, finite-element methods (FEM) provide a cost-effective alternative for performing preliminary feasibility analyses of various geometric and material parameters. FEM simulations allow for the assessment of thermomechanical stresses, temperature distributions, and the influence of different material properties on interconnect performance, which can greatly reduce the need for physical prototypes and experiments. Despite this, FEM studies are often limited by the complexity of the interconnect structure, particularly when dealing with the substantial size disparities across different layers of a TSV interconnect. For example, the silicon substrate can span hundreds of micrometers, while TSV-Cu and the RDL circuitry can have much smaller, sub-micron scale features. This substantial difference in dimensional scales makes mesh generation a challenging task, resulting in significant computational difficulties. As a result, many studies fail to consider the intricate details of the RDL layer, or they make simplifying assumptions that limit the accuracy of the analysis [3–10].

Moreover, understanding the interaction between TSV and RDL becomes even more critical once cracks have already formed at the interfaces. The complex behavior of interfacial cracks, combined with the residual stresses and varying loading conditions, necessitates a more advanced approach to modeling. Fracture mechanics-based FEM provides an effective solution to this problem, allowing for the prediction of cracking risks by calculating fracture mechanics parameters—such as the energy release rate, stress intensity factors, or J-integrals—at the crack tips [11–13]. By comparing these parameters with critical values, the likelihood of structural failure under different packaging loads can be assessed. This approach has the advantage of not requiring detailed material fracture criteria inputs, making it highly efficient and practical for predicting failure risks.

In this chapter, the use of fracture mechanics modeling to investigate the effects of residual stress, geometric parameters, and thermodynamic conditions on cracking behavior in TSV/RDL interconnect structures is emphasized. By simulating the behavior of these interconnects under annealing processes, the study provides important insights into how these factors influence the mechanical integrity of interconnects. The results can serve as a valuable reference for the design of more reliable TSV interconnect structures and contribute to improving the overall robustness of advanced electronic packaging. This approach not only addresses the limitations of traditional methods but also opens new avenues for designing more resilient interconnect systems in high-performance integrated circuits.

7.2   Interface fracture mechanics theory

To evaluate the interaction effects between TSV and RDL using fracture mechanics methodologies, it is necessary to introduce fracture mechanics parameters such as the energy release rate and the J-integral. For a two-dimensional interfacial crack involving two linear isotropic materials, as illustrated in Figure 7.1, the expression for the stress field at the crack tip is as follows [14]:

$$\sigma_{yy} + i\sigma_{xy} = \frac{K}{\sqrt{2\pi r}} r^{i\varepsilon} \tag{7.1}$$

where r represents the distance from a point near the crack to the crack tip in Figure 7.1. $K = K_t + iK_2$ denotes the complex stress intensity factor. The constant

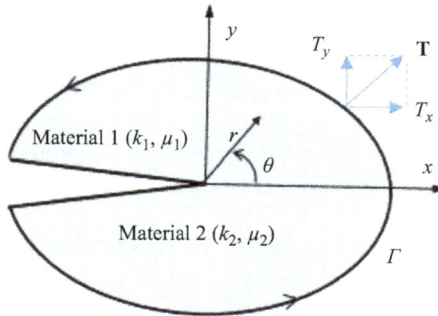

Figure 7.1   Crack on the interface between two dissimilar materials

ε for the two-dimensional interfacial bi-material is defined as follows:

$$\varepsilon = \frac{1}{2\pi} \ln\left[\left(\frac{k_1}{\mu_1} + \frac{1}{\mu_2}\right) \Big/ \left(\frac{k_2}{\mu_2} + \frac{1}{\mu_1}\right)\right] \tag{7.2}$$

where subscripts 1 and 2 denote the two bonded materials at the interface. Based on the two-dimensional plane strain or plane stress model, the expression for $k_\alpha(\alpha = 1, 2)$ is as follows:

$$k_\alpha = \begin{cases} 3 - 4\nu_\alpha \\ (3 - \nu_\alpha)/(1 + \nu_\alpha) \end{cases} \tag{7.3}$$

where ν_α and μ_α represent Poisson's ratio and shear modulus, respectively. The expression for the crack opening displacement at a distance r from the crack tip is as follows:

$$\delta_y + i\delta_x = \frac{c_1 + c_2}{\sqrt{2\pi r}(1 + 2i\varepsilon) \cosh(\pi\varepsilon)} Kr^{i\varepsilon}\sqrt{r} \tag{7.4}$$

where the expression for c_1 and c_2 are

$$\begin{cases} c_1 = \dfrac{k_1 + 1}{\mu_1} \\ c_2 = \dfrac{k_2 + 1}{\mu_2} \end{cases} \tag{7.5}$$

The expression for the energy release rate of a two-dimensional interfacial crack surface is as follows:

$$G = \frac{(c_1 + c_2)K\bar{K}}{16 \cosh^2(\pi\varepsilon)} \tag{7.6}$$

where $\bar{K}$ is the conjugate complex form of K.

The aforementioned discussion presents the expression for the energy release rate, applicable to linear elastic fracture mechanics. When plastic deformation occurs in the vicinity of the crack tip, the J-integral must be used to describe the energy field at the crack tip. According to studies [15,16], the expression of the J-integral at the crack tip of a two-dimensional interfacial crack in bi-materials is similar to the standard J-integral expression for cracks at the tips of homogeneous material interfaces, as shown in the following equation:

$$J = \int_\Gamma \left[Wdy - T_i\frac{du_i}{dx}ds\right] \tag{7.7}$$

where Γ represents an arbitrary closed counterclockwise integral contour path surrounding the crack tip in the Cartesian coordinate system, W is the strain energy density per unit volume, and ds is the infinitesimal integral element of the arc length of the integral contour. T_i and u_i are the traction and displacement vectors of the integral contour, respectively.

7.3   Modeling method of TSV/RDL interconnect structures with preset cracks

7.3.1   *Geometric model and material parameters*

The TSV/RDL interconnect structure studied in this section can be fabricated using the Via last process, with the main fabrication steps illustrated in Figure 7.2. After preparing the complementary metal-oxide-semiconductor (CMOS) device layer and the RDL dielectric layer on the Si wafer, the subsequent TSV fabrication process includes etching blind holes, depositing an SiO_2 insulating layer, copper electroplating, chemical mechanical polishing, and annealing. To investigate the interaction between TSV-Cu and the RDL layer during the annealing process and to optimize the interconnect structure using fracture mechanics methods, a two-dimensional local finite-element model of the TSV wafer was constructed using the finite-element software ABAQUS, as shown in Figure 7.3(a). The model is based on the plane strain assumption, with symmetric constraints applied along the y-direction at the left and right edges of the model, and a fixed constraint applied to the bottom left node to prevent rigid body motion. The total thickness w of the Si wafer is 500 μm, with a width of $P/2$, where P is the pitch of the TSV. A total thickness of 5.5 μm RDL layer, including a 0.5 μm thick SiO_2 insulating layer, is established on the front side of the wafer. It is important to note that the model shown in Figure 7.3(a) is a simplified representation of the TSV/RDL interconnect structure, with the RDL layer constructed according to the dimensions of the actual scanning electron microscope (SEM) image

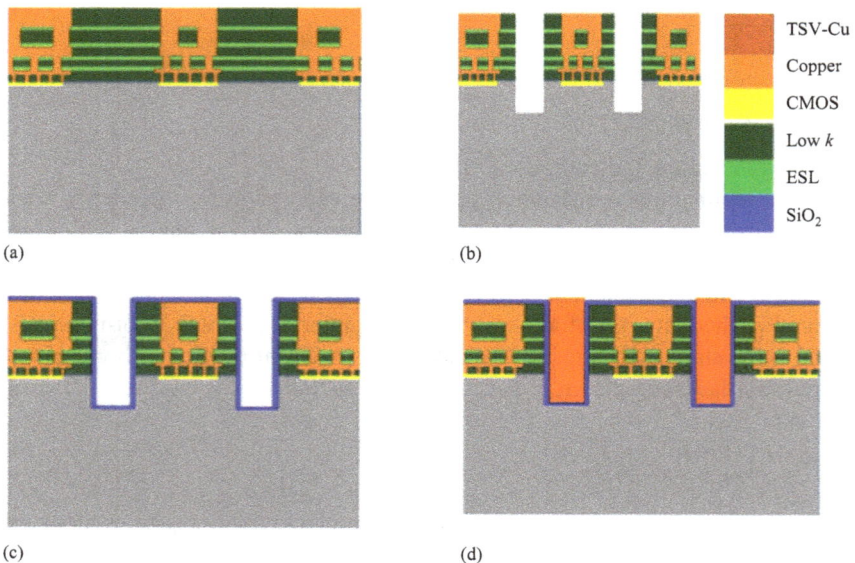

Figure 7.2   The main steps of manufacturing TSV/RDL interconnected structure by the via last process: (a) the manufactured CMOS devices and RDL layers, (b) etch TSV blind via, (c) deposit SiO_2 insulator, and (d) electroplating of TSV-Cu, followed by CMP and annealing

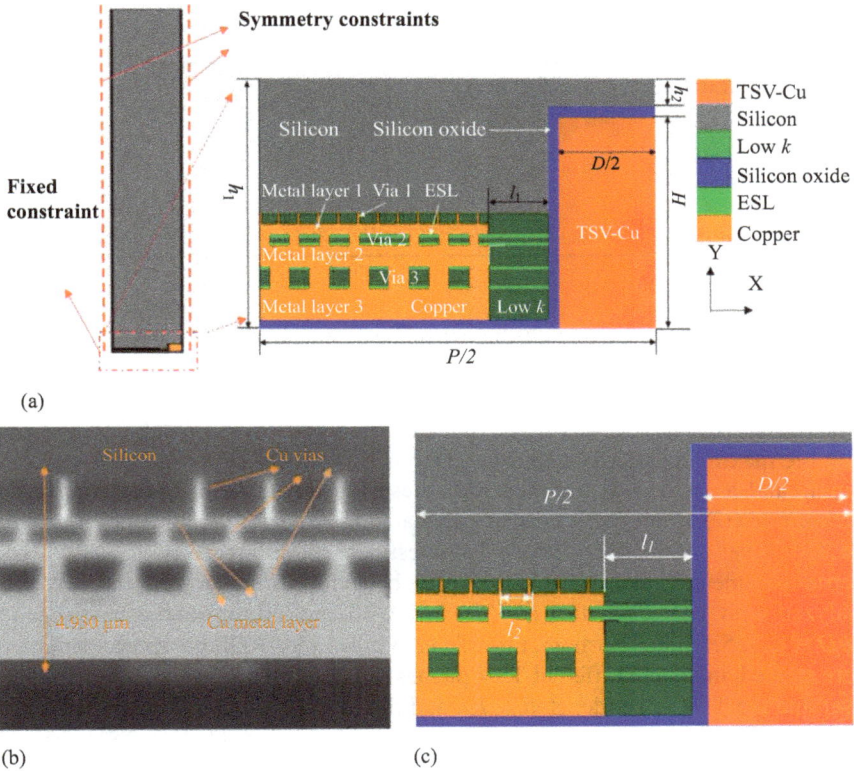

(a)

(b)   (c)

Figure 7.3   Schematic diagram of finite-element model of TSV wafer: (a) the two-dimensional local finite element model of TSV wafer, (b) SEM image of the RDL layers, and (c) geometric parameters in the model

Table 7.1   Dimension of Cu via and low k interconnection structure

Layers	Thickness (µm)	Width (µm)	Pitch (µm)
Metal layer 1	0.5	–	–
Metal layer 2	1	–	–
Metal layer 3	1.5	–	–
Via 1	0.5	0.1	1
Via 2	0.5	1	1
Via 3	1	1	1
ESL 1	0.1	–	–
ESL 2	0.1	–	–

shown in Figure 7.3(b). The thickness of the Cu metal layer gradually decreases from the third metal layer to the first metal layer. The specific dimensions of the RDL layer, which interconnects the Cu metal layer, low-k dielectric layer, and etch stop layer (ESL), are shown in Table 7.1. A temperature load is applied to the model to simulate

Table 7.2   Material properties of MSQ [17]

Materials	E (GPa)	Poisson's ratio	α
MSQ-A	2	0.3	4
MSQ-B	4	0.3	10
MSQ-C	7.7	0.3	18
MSQ-D	10	0.3	23
MSQ-E	17	0.3	26

the TSV annealing process, with the annealing temperature set at 400 °C, followed by cooling to room temperature (25 °C), with a reference temperature of 400 °C.

To assess the most vulnerable locations in the TSV/RDL interconnect model, an annealing simulation was first conducted on a crack-free model. The positions with the maximum principal stress and strain on the crack-free model were determined, and a crack was pre-placed at these locations. Additionally, the energy release rate (G) at the crack tip was calculated to evaluate the interaction between the TSV layer and the RDL layer. To investigate the influence of TSV/RDL geometric parameters on interfacial fracture behavior, finite-element models were analyzed with TSV diameters of (D = 5, 10, 30, and 50 μm) and aspect ratios of (H/D = 1, 3, 5, 7). Furthermore, the effects of other TSV/RDL geometric parameters were studied, including different TSV pitches (P), the gap distance between the RDL layer and the TSV edge (denoted as l_1), and the pitch of Cu hole 2 in the RDL layer (denoted as l_2). The schematic representation of Cu hole 1, Cu hole 2, TSV pitch (P), and the distance l_1 between TSV-Cu and RDL is shown in Figure 7.3(c).

In addition to investigating the influence of the aforementioned TSV/RDL geometric parameters on interfacial cracking behavior, further exploration will be conducted on the thermodynamic parameters of TSV-Cu and RDL. Various porous low-k dielectric materials have been developed in existing literature to enhance the overall performance of the RDL layer. To examine the effects of different low-k dielectric materials (MSQ) on interfacial fracture behavior, the following subsection discusses the impact of the elastic modulus and thermal expansion coefficient (α) of different dielectric materials on the energy release rate of interfacial cracks, as illustrated in Table 7.2. In the aforementioned finite-element model analyses, Cu material is assumed to be linearly elastic, as a significant body of existing literature on the reliability numerical analysis of Cu interconnect structures is based on the assumption that Cu behaves as a linear elastic material. The subsequent subsection also considers the plastic material parameters of Cu, as illustrated in Table 7.3, to study the differences in the impact of linear elastic parameters versus elastoplastic parameters of Cu on the energy release rate of interfacial cracks.

7.3.2   Finite-element mesh and crack location

The analysis results of the crack-free model are illustrated in Figure 7.4, where the TSV diameter D is set at 10 μm, the pitch P is 20 μm, and the aspect ratio H/D is 1.

Table 7.3   Electroplated copper material properties [18]

E (GPa)	Poisson's ratio	α (ppm/°C)	Plastic properties (MPa)
70	0.34	17.6	240 @ 0 ε
			250 @ 0.003 ε
			244 @ 0.007 ε
			255 @ 0.009 ε
			250 @ 0.017 ε

(a)

(b)

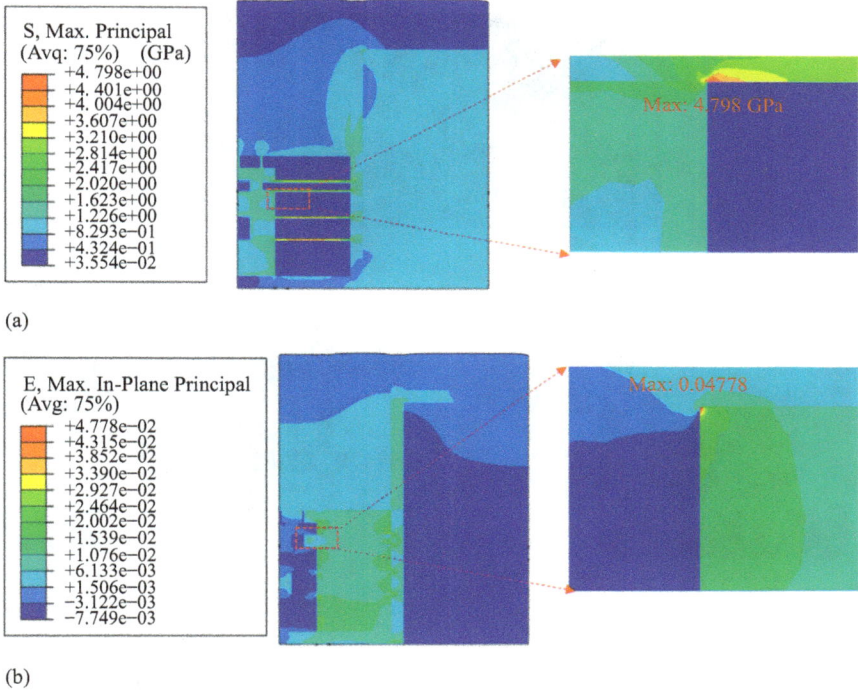

Figure 7.4   Contours of max principal stress and principal strain for the model: (a) max principal stress contour and (b) max principal strain contour

The material parameters of the model, all of which are considered linear elastic materials, are listed in Table 7.4. The maximum principal stress and strain are measured at 4.798 GPa and 0.0478, respectively. Furthermore, the locations of the maximum principal stress and strain are found at the corner at the interface between the second metal layer and the ESL, indicating that this corner point is prone to failure. Consequently, in the subsequent finite-element analyses, an L-shaped crack will be pre-placed at this corner point to calculate the fracture mechanics parameters at the crack tip under various influencing conditions.

Table 7.4   Material properties used in the FEA

Materials	E (GPa)	α (ppm/$^\circ$C)	Poisson's ratio
Silicon	130	2.8	0.28
Silicon oxide	71	1	0.17
TSV-Cu	110	17	0.3
Copper	121	17.6	0.3
Low k	7.7	23	0.3
ESL (SiN)	310	3	0.27

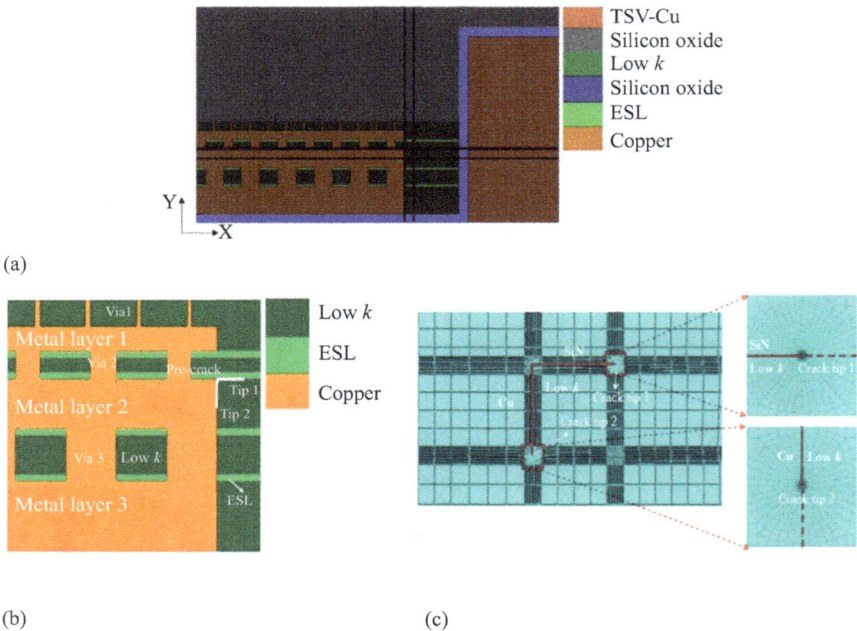

(a)

(b)   (c)

Figure 7.5   Schematic diagram of finite-element meshes: (a) finite element meshes of TSV and RDL structure, (b) the position of the L shaped crack in the RDL structure, and (c) finite element meshes of local L shaped crack region

The finite-element mesh of the pre-placed crack model is illustrated in Figure 7.5(a), where the element type used is a bilinear plane strain quadrilateral four-node reduced integration element (CPE4R) with an element length of 0.1 μm. The total number of elements in the finite-element models under different influencing parameters varies between 90,084 and 199,148. Additionally, the position of the L-shaped crack in the model is shown in Figure 7.5(b), which interfaces two boundaries: the ESL/low-k dielectric material interface and the Cu/low-k dielectric material interface. The horizontal and vertical lengths of the L-shaped crack are both set to 0.5 µm. The edge lengths of the finite-element units in the localized L-shaped crack region range from 0.001 to 0.1 µm, as depicted in Figure 7.5(c).

This study directly extracts the energy release rate G at the two crack tips of the L-shaped crack from ABAQUS. The G values are obtained from ten contour integrals in the crack tip region. To enhance accuracy, the G value computed from the first contour integral is disregarded. The final G value is thus calculated as the average of the subsequent nine contour integrals.

7.4 Factors influencing the cracking behavior of TSV/RDL interconnect structures

Residual stress and interfacial cracking are two critical factors that affect the reliability of TSV interconnections and the surrounding devices. This subsection discusses the impact of several parameters, including the diameter (D) of the TSV, the pitch (P), the separation distance (l_1) between the TSV and the RDL layer, the spacing (l_2) of the second layer Cu vias within the RDL layer, as well as the material properties of the TSV-Cu and low-k dielectric on the cracking behavior of the TSV/RDL interconnection structure.

7.4.1 Residual stress in TSV/RDL interconnection structures

To analyze the distribution of residual stress in the TSV and RDL layers near the silicon surface during the annealing process, a finite-element model with a TSV diameter D of 10 µm, a pitch P of 20 µm, and an aspect ratio H/D of 1 was employed to calculate the residual stress distribution post-annealing. In the model, a vertical path Y is defined from the top center point of the TSV-Cu to its bottom center point, as indicated by the red dashed line in Figure 7.6. The stress components (σ_{22}, σ_{11}, and σ_{12}) along path Y were directly extracted from ABAQUS. As shown in Figure 7.6(a), the axial stress σ_{22} in the TSV is tensile, gradually decreasing from approximately 900 MPa at the bottom of the TSV-Cu to 200 MPa in the downward vertical direction. The radial stress σ_{11} and shear stress σ_{12} are relatively small compared to the axial stress σ_{22}, remaining roughly within the range of 100 MPa; thus, the variations of radial stress σ_{11} and shear stress σ_{12} will be presented within a smaller stress range. From Figure 7.6(b), it is evident that the radial stress σ_{11} is tensile, initially increasing slowly from about 80 MPa as the distance from the TSV top increases, reaching a maximum value of 106 MPa at the upper edge of the RDL layer, followed by stabilization, and then gradually decreasing to zero at the bottom of the TSV. Figure 7.6(c) clearly illustrates that the shear stress σ_{12} initially registers at -45 MPa, gradually diminishing to zero near the upper edge of the RDL layer, before increasing in the opposite direction, achieving a maximum value of 60 MPa at the edge of the second metal layer, and then decreasing back to zero.

To further analyze the distribution of residual stress within the RDL layer, four horizontal paths, denoted as X-path 1, X-path 2, X-path 3, and X-path 4, were defined, as illustrated in Figure 7.7(a). Specifically, X-path 1 traverses the Si

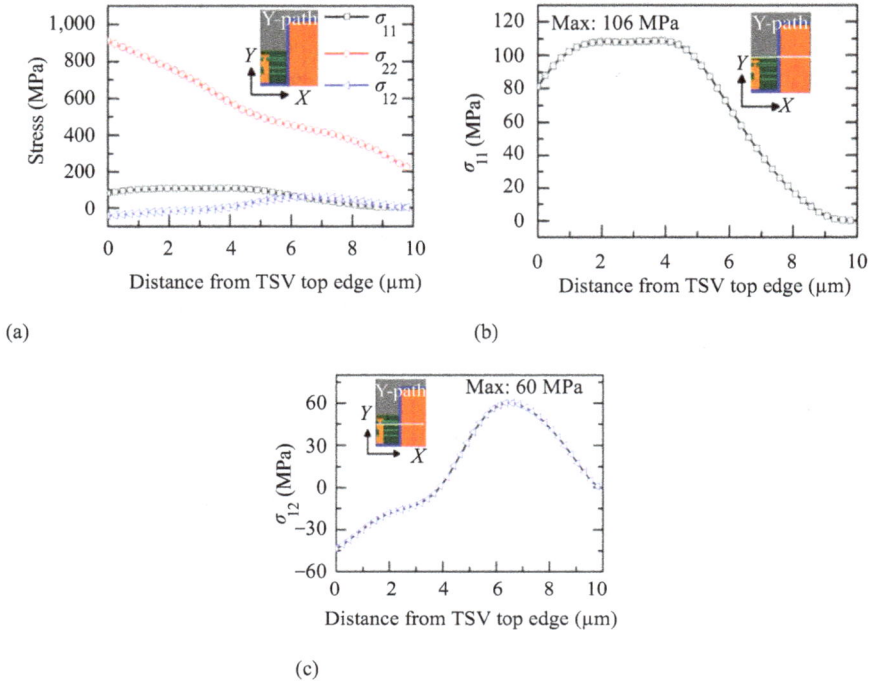

(a)

(b)

(c)

*Figure 7.6   Residual stress distributions along the vertical Y-path in TSV:
(a) stress components along the vertical Y- path, (b) radial stress σ_{11},
and (c) axial stress σ_{22}*

substrate and the "TSV-Cu zone," while X-path 2, X-path 3, and X-path 4 extend through the "TSV-Cu zone," "low-k zone," and "Cu zone," respectively. Figure 7.7 (b) presents the axial stress σ_{22} as it varies with the distance from the right end (TSV-Cu zone) to the left end along these four X-paths.

The results indicate that within the range of 0–8 μm, σ_{22} along X-path 1 is relatively high, and the residual stress in the Si substrate is greater than that inside the TSV-Cu. The stress range in Si varies between 600 and 900 MPa, while the stress range in TSV-Cu is between 400 and 500 MPa. Additionally, the range of σ_{22} in the "TSV-Cu zone" and "Cu zone" is 300–500 MPa and 400–900 MPa, respectively, gradually decreasing from X-path 2 to X-path 4. However, σ_{22} remains approximately equal in the "low-k zone," at around 200 MPa. Radial stress σ_{11}, as shown in Figure 7.7(c), is tensile along X-path 1 in the "TSV-Cu zone" and compressive in the "low-k zone" and "Cu zone." Nevertheless, σ_{11} remains almost equal along the other three paths. It is worth noting that σ_{12} experiences a sudden drop on all paths when passing through SiO_2. Shear stress σ_{12}, as illustrated in Figure 7.7(c), is zero at both ends of the four paths. σ_{12} remains negative along X-path 1, gradually increasing in the "TSV-Cu zone" and decreasing in the Si substrate. The variation pattern of σ_{12} is

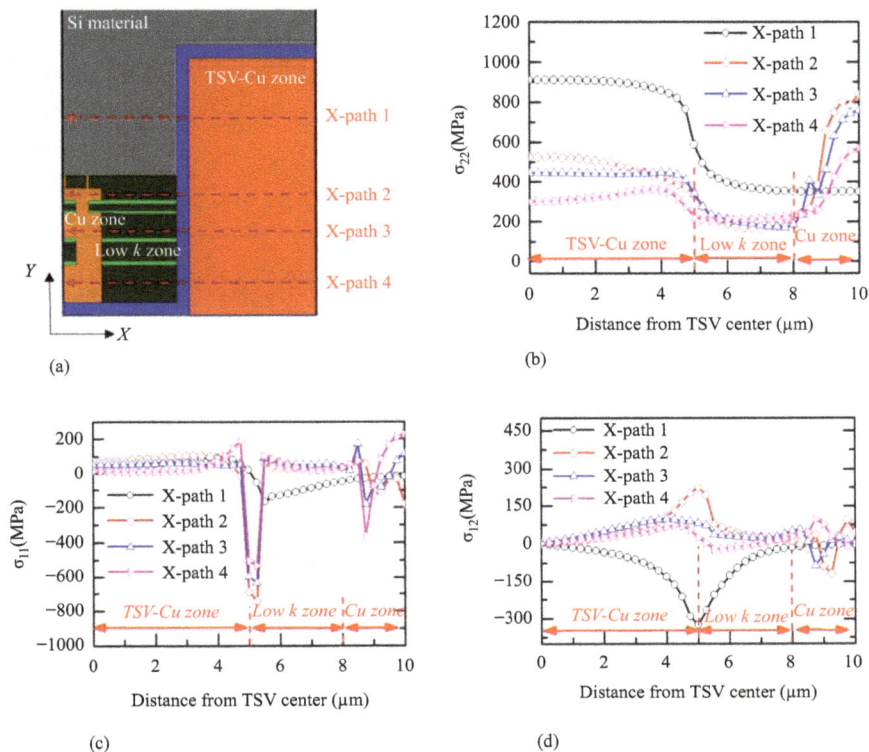

Figure 7.7   Comparison of residual stress components along four horizontal paths: (a) schematic of four horizontal paths, (b) comparison of axial stress σ₂₂, (c) comparison of radial stress σ₁₁, and (d) comparison of shear stress σ₁₂

similar along the other three paths, increasing initially in the "TSV-Cu zone" and decreasing subsequently in the "low-k zone" and "Cu zone." The distribution of residual stress within the TSV and RDL layers demonstrates the interactive effects between the TSV and RDL layers. The presence of TSV-Cu influences the residual stress distribution in the RDL layers and provides a driving force for the formation of L-shaped cracks within the RDL layers.

7.4.2   Geometric parameters of TSV/RDL

Incrementally reducing the diameter and pitch of TSV enhances the interconnect density of TSV-Cu and strengthens the interaction between the TSV and RDL layers. This section investigates the influence of geometric parameters of the TSV/RDL layers on the energy release rate at the tip of L-shaped interfacial cracks. Finite-element models were established with TSV diameters of 50, 30, 10, and 5 μm, as well as various aspect ratios ($H/D = 1, 3, 5, 7$). The total length of the

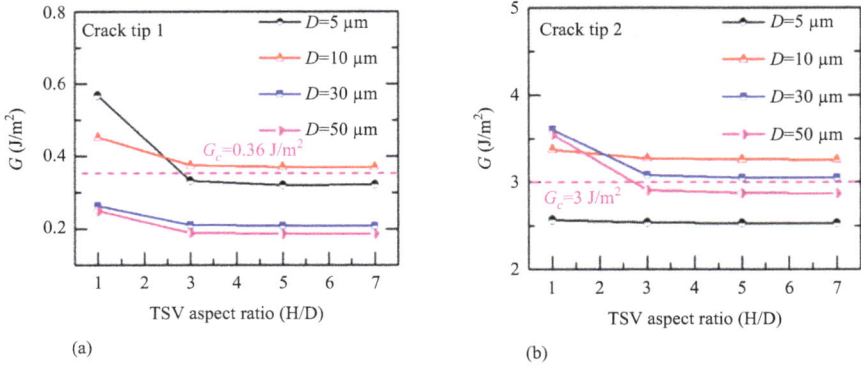

Figure 7.8   Energy release rate of crack tips with different TSV's aspect ratios:
(a) crack tip 1 energy release rate and (b) crack tip 2 energy release rate

model was 2.5D, with a fixed distance l_1 (l_1 = 3 µm). Research indicates that the average value of the elastic modulus of low-k dielectric materials is 7 GPa, corresponding to a critical energy release rate of 0.36 J/m^2, which is used as the critical energy release rate for low-k dielectric materials [19]. The energy release rate for the Cu/low-k dielectric material interface ranges from 3 to 5 J/m^2 [17–20], with a minimum value of 3 J/m^2 assumed as the critical energy release rate G_c for interfacial cracks in the Cu/low-k material in subsequent calculations.

The variation of the energy release rate at the crack tips with the TSV aspect ratio (H/D) is shown in Figure 7.8. The results demonstrate that the energy release rate (G) at crack tip 1, composed of the ESL/low-k dielectric material interface, initially decreases with increasing aspect ratio (H/D) before stabilizing, as shown in Figure 7.8(a). Similarly, G at crack tip 2, composed of the Cu/low-k dielectric material interface, also decreases initially and then stabilizes. G at crack tip 2 is generally an order of magnitude larger than that at crack tip 1, indicating a greater propensity for interfacial cracking behavior in the Cu/low-k dielectric material interface. Additionally, as the TSV diameter increases and the aspect ratio (H/D) is greater than or equal to 3, G at both crack tip 1 and crack tip 2 initially increases and then decreases. When the aspect ratio (H/D) is 3, G at both crack tips under the same TSV diameter stabilizes. In summary, larger aspect ratios contribute to a reduction in the energy release rate.

To investigate the influence of TSV pitch on the energy release rate of L-shaped interfacial cracks, five finite-element models with different TSV pitches (P/D) were established, specifically at ratios of 2, 3, 4, 6, and 8. The diameter, height, and distance (l_1) of the TSV were fixed at 10, 10, and 3 µm, respectively. The energy release rate (G) at the tips of L-shaped interfacial cracks with different TSV pitches is illustrated in Figure 7.8. The results indicate that the G at both crack tip 1 (shown in Figure 7.9(a)) and crack tip 2 (shown in Figure 7.9(b)), increases with the increase in pitch. However,

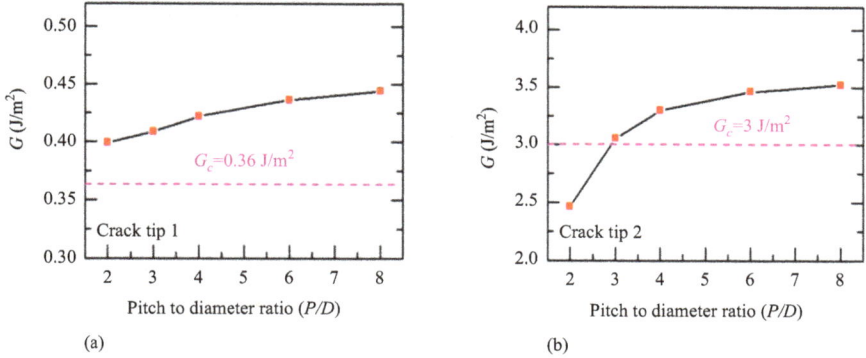

Figure 7.9 *Energy release rate of crack tips with different TSV's pitches: (a) crack tip 1 energy release rate and (b) crack tip 2 energy release rate*

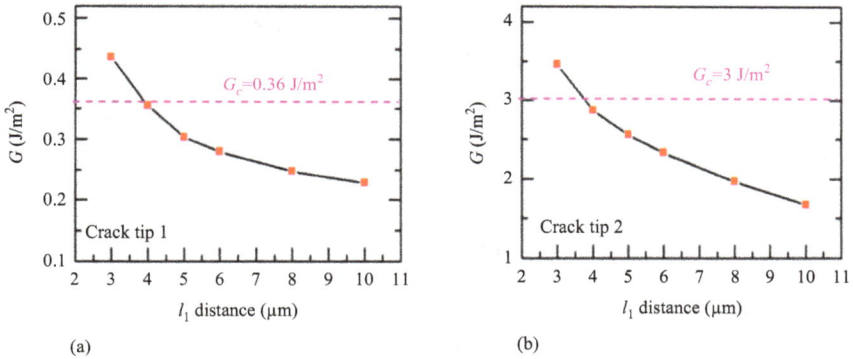

Figure 7.10 *Energy release rate of crack tips with different l_1 distances: (a) crack tip 1 energy release rate and (b) crack tip 2 energy release rate*

as the TSV pitch increases, G surpasses the critical energy release rate G_c, suggesting that smaller TSV pitches are beneficial for reducing the risk of interfacial fracture.

To analyze the impact of the distance (l_1) between the RDL and the TSV edge, as well as the spacing (l_2) between copper vias on interfacial cracking, five finite-element models with varying l_1 distances (l_1= 3, 4, 5, 6, 8 μm) and two different Cu via spacings (l_2= 1 μm, l_2= 0.5 μm) were established. The TSV diameter (D), TSV pitch (P/D), and Cu via spacing (l_2) were maintained constant at 10, 2, and 1 μm, respectively. The energy release rates at both tips of L-shaped interfacial cracks under different l_1 distances are depicted in Figure 7.10. The results indicate that as l_1 increases, the G values at both crack tips exhibit a gradual decrease, falling below the critical energy release rate. Specifically, the G at crack tip 1, as shown in Figure 7.10(a), decreases from approximately 0.45 to around 0.22 J/m^2, while the G at crack tip 2, as shown in Figure 7.10(b), declines from about 3.5 to approximately 1.6 J/m^2.

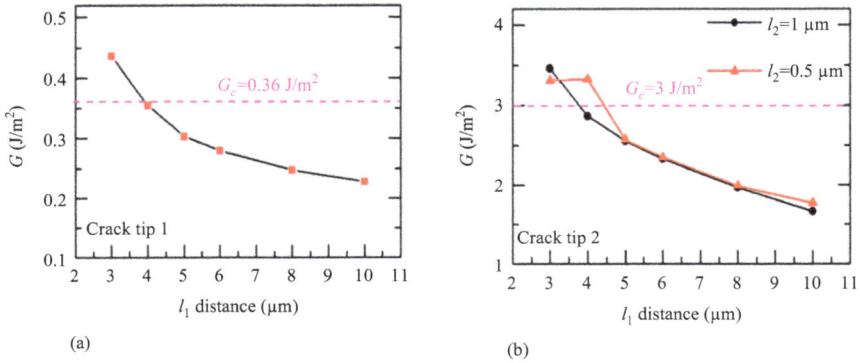

Figure 7.11   Energy release rate of crack tips with different l_2 distances: (a) crack tip 1 energy release rate and (b) crack tip 2 energy release rate

The energy release rates at the tips of L-shaped interfacial cracks for different Cu via spacings are illustrated in Figure 7.11. The findings reveal that the G values for various l_2 spacings are nearly equivalent. However, when l_1 is set to 4 µm, the G of the model with a Cu via spacing of 1 µm is less than that of the model with l_2= 0.5 µm. This suggests that at l_1= 4 µm, the model with a wider Cu via spacing of 1 µm exhibits a lower energy release rate. With the aim of fully utilizing space and integrating more devices without exceeding G_c, a relatively smaller distance l_1 can suffice when l_2 is configured at 1 µm.

7.4.3   Thermodynamic parameters of the RDL

This section discusses the influence of thermodynamic parameters of low-k dielectric materials on the cracking behavior of interfacial cracks. According to the various porous low-k dielectric materials listed in Table 7.2, specifically MSQ (A–E), their elastic moduli E range from 2 to 17 GPa, while their coefficient of thermal expansion (CTE) vary from 4 to 26 ppm/°C, encompassing the variation ranges of E and CTE for most dielectric materials currently available. During the modeling process, the diameter of the TSV (D), aspect ratio (H/D), TSV spacing (P/D), and the spacing of copper vias (l_2) were kept constant at 10, 1, 2, and 1 µm, respectively. The results, illustrated in Figure 7.12, indicate that as E increases, the G values at the two crack tips increase significantly, even exceeding the critical energy release rate G. Conversely, when E is held constant, an increase in l_1 leads to a gradual decrease in G as shown by the green dashed line in Figure 7.12.

However, as the CTE increases, the G value at crack tip 1 initially decreases slowly before gradually increasing, illustrated in Figure 7.13(a); in contrast, the G value at crack tip 2 monotonically increases with the increase of CTE, illustrated in Figure 7.13(b). For a fixed CTE (indicated by the green dashed line in Figure 7.13), the G values generally decrease with the increasing distance l_1. Notably, when the CTE is less than 18 ppm/°C (circled with a red dashed line in Figure 7.13(a)), the G

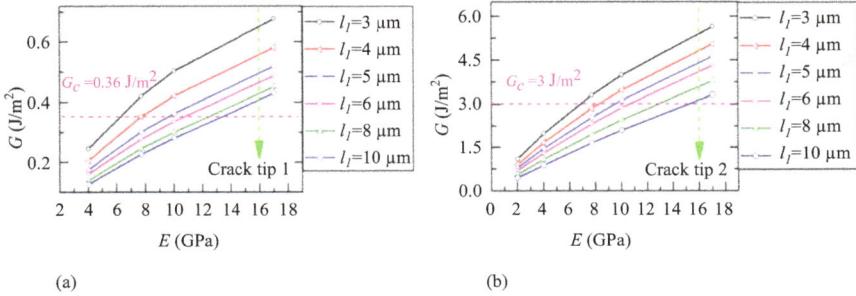

Figure 7.12 *Variation of energy release rate of crack tips with E under different l_2 distances: (a) energy release rate of crack tip 1 and (b) energy release rate of crack tip 2*

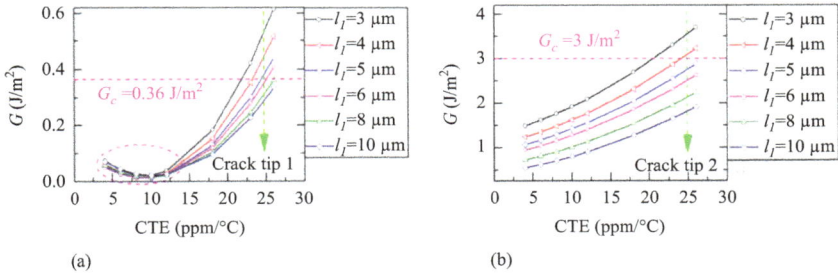

Figure 7.13 *Variation of energy release rate of crack tips with CTE under different l_1 distances: (a) energy release rate of crack tip 1 and (b) energy release rate of crack tip 2*

values at crack tip 1 remain approximately equal for a given CTE, regardless of the differing l_1 distances. It is important to emphasize that a smaller l_1 distance allows for the integration of more devices while maintaining a high reliability of the low-*k* dielectric materials' interconnections. Based on the comparison between the calculated energy release rates at the crack tips and the critical energy release rate, selecting dielectric materials with an *E* value below 10 GPa and a CTE value below 23 ppm/°C can mitigate the risk of interfacial cracking in the RDL.

7.4.4 Mechanical properties of TSV-Cu

Due to the annealing simulation temperature cooling from 400 °C to room temperature (25 °C), such high-temperature annealing conditions may lead to plastic deformation of the copper in the TSV-Cu and RDL layers, consequently affecting the reliability of the TSV/RDL interconnection layer. Most existing numerical studies on the reliability of Cu interconnection structures assume Cu to be a linearly elastic material, which may not yield high accuracy [21,22]. To investigate the differences in the effects of linear elastic parameters and elastoplastic parameters of

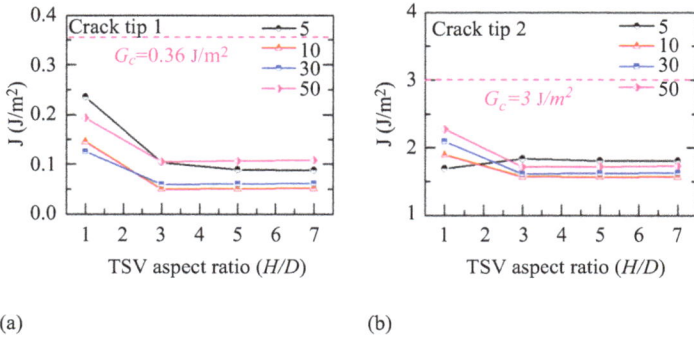

Figure 7.14   *Energy release rate of crack tips under the elastoplastic model: (a) crack tip 1 energy release rate and (b) crack tip 2 energy release rate*

Cu on interfacial cracking, the Cu material in the model adopts the elastoplastic parameters listed in Table 7.3. The criterion for the energy release rate in the framework of fracture mechanics is only applicable to linear elastic fracture mechanics. Therefore, when plastic deformation occurs, the J-integral criterion is utilized to analyze the impact of plastic deformation on the cracking behavior at the tips of L-shaped interfacial cracks.

As illustrated in Figure 7.14, the stress relaxation due to the plastic deformation of Cu results in a J-integral that is generally lower for the elastoplastic model compared to the energy release rate of the linear elastic model. Additionally, the J-integrals at both crack tips are less than the temporary energy release rate G_c, indicating that the plastic deformation of Cu induced by the annealing process reduces the risk of interfacial cracking in the interconnection structure. For the same TSV diameter, the J-integral initially decreases as the TSV aspect ratio (H/D) increases, stabilizing when the aspect ratio is greater than or equal to 3. However, when the TSV diameter $D=5$ μm, the J-integral at crack tip 2 first increases before stabilizing, illustrated in Figure 7.14(b). The results show that the J-integral value at crack tip 1 for the elastoplastic model ranges from 0.05 to 0.25 J/m^2, while the J-integral value at crack tip 2 ranges from 1.5 to 2.5 J/m^2. In contrast, under the same geometric conditions in Section 7.4.2, the energy release rate values for crack tip 1 of the linear elastic model range from 0.2 to 0.6 J/m^2, and the energy release rate values for crack tip 2 range from 2 to 3.5 J/m^2. It is evident that, compared to the linear elastic model, the elastoplastic model helps to reduce the risk of cracking in the RDL layer.

7.5   Chapter summary

This chapter presents a modeling approach rooted in fracture mechanics to simulate the residual stress distribution in TSV and RDL interconnect structures following the annealing process. The primary focus is to investigate how various geometric and thermomechanical parameters of TSV/RDL structures influence interfacial cracking behavior, which is critical for the reliability and longevity of advanced packaging technologies.

The study reveals several important findings regarding the influence of TSV geometry on the interconnect reliability. As the aspect ratio (H/D, where H is the TSV height and D is the TSV diameter) increases, the energy release rate initially decreases. This rate stabilizes and remains relatively constant for aspect ratios greater than or equal to 3. A higher aspect ratio and smaller TSV pitches (distance between vias) effectively reduce the energy release rate at the interface, thereby mitigating the risk of interfacial cracking. The results suggest that an optimized TSV structure, with high aspect ratios and tight pitches, can significantly reduce stress concentrations and improve the interconnect's robustness.

Moreover, the distance (l1) between the TSV and the RDL layer plays a crucial role in controlling the stress at the interface. When the distance is increased to 4 μm, and the Cu via pitch (l2) is reduced to 1 μm, the energy release rate drops below the critical threshold, suggesting that these dimensions are conducive to reducing cracking risks while maintaining a high device density. This finding is valuable for optimizing the TSV/RDL design to meet both reliability requirements and integration constraints.

The study also emphasizes the importance of low-k dielectric materials in influencing interfacial cracking behavior. Materials with a Young's modulus lower than 10 GPa and a CTE below 23 ppm/°C exhibit a reduced risk of stress-induced cracking. These properties are beneficial for minimizing the mismatch in thermal expansion between the different layers of the interconnects, thereby improving the overall mechanical stability.

Additionally, the plastic deformation of both the TSV-Cu and RDL materials is considered in the analysis. It is found that plastic deformation helps to absorb and redistribute a portion of the stress, which further reduces the likelihood of interfacial cracking. This plastic behavior contributes to stress relaxation, thereby enhancing the reliability of the interconnect structure.

Overall, the results presented in this study offer valuable insights for the design and reliability assessment of TSV-based packaging structures, helping to optimize geometric parameters and material choices to minimize the risk of failure due to interfacial cracking.

References

[1] K.H. Lu, S.-K. Ryu, Q. Zhao, *et al.*, Thermal stress induced delamination of through silicon vias in 3-D interconnects, in: *2010 Proceedings 60th Electronic Components and Technology Conference (ECTC)*, IEEE; 2010: pp. 40–45.

[2] L. Spinella, T. Jiang, N. Tamura, J.-H. Im, and P.S. Ho, Synchrotron X-ray microdiffraction investigation of scaling effects on reliability for through-

silicon vias for 3-D Integration, *IEEE Transactions on Device and Materials Reliability* 19 (2019) 568–571.

[3]   I. De Wolf, K. Croes, O. Varela Pedreira, *et al.*, Cu pumping in TSVs: Effect of pre-CMP thermal budget, *Microelectronics Reliability* 51 (2011) 1856–1859.

[4]   S.-K. Ryu, Q. Zhao, M. Hecker, *et al.*, Micro-Raman spectroscopy and analysis of near-surface stresses in silicon around through-silicon vias for three-dimensional interconnects, *Journal of Applied Physics* 111 (2012) 063513.

[5]   P. Dixit, S. Yaofeng, J. Miao, J.H.L. Pang, R. Chatterjee, and R.R. Tummala, Numerical and experimental investigation of thermomechanical deformation in high-aspect-ratio electroplated through-silicon vias, *Journal of the Electrochemical Society* 155 (2008) H981.

[6]   H. Kim, H. Jeon, D.-J. Lee, and J.-Y. Kim, Surface residual stress in amorphous SiO_2 insulating layer on Si substrate near a Cu through-silicon via (TSV) investigated by nanoindentation, *Materials Science in Semiconductor Processing* 135 (2021) 106153.

[7]   P. Saettler, M. Hecker, M. Boettcher, C. Rudolph, and K.J. Wolter, μ-Raman spectroscopy and FE-modeling for TSV-stress-characterization, *Microelectronic Engineering* 137 (2015) 105–110.

[8]   M.K. Yah and C.L. Lu, Reliability analysis of 3D heterogeneous microsystem module by simplified finite element model, *Microelectronics Reliability* 63 (2016) 111–119.

[9]   S.-K. Ryu, K.-H. Lu, X. Zhang, J.-H. Im, P.S. Ho, and R. Huang, Impact of near-surface thermal stresses on interfacial reliability of through-silicon vias for 3-D Interconnects, *IEEE Transactions on Device and Materials Reliability* 11 (2011) 35–43.

[10] I.H. Jeong, M.H. Roh, F. Jung, W.H. Song, M. Mayer, and J.P. Jung, Analysis of the electrical characteristics and structure of Cu-filled TSV with thermal shock test, *Electronic Materials Letters* 10 (2014) 649–653.

[11] X.H. Liu, M.W. Lane, T.M. Shaw, and E. Simonyi, Delamination in patterned films, *International Journal of Solids and Structures* 44 (2007) 1706–1718.

[12] L.L. Mercado, C. Goldberg, S.-M. Kuo, Tien-Yu, T. Lee, and S.K. Pozder, Analysis of flip-chip packaging challenges on copper/low-*k* interconnects, *IEEE Transactions on Device and Materials Reliability* 3 (2003) 111–118.

[13] J. Auersperg, D. Vogel, E. Auerswald, S. Rzepka, and B. Michel, Nonlinear copper behavior of TSV and the cracking risks during BEoL-built-up for 3D-IC-integration, in: *2012 13th International Thermal, Mechanical and Multi-Physics Simulation and Experiments in Microelectronics and Microsystems*, IEEE; 2012: p. 1/6–6/6.

[14] J.R. Rice, Elastic fracture mechanics concepts for interfacial cracks, *Journal of Applied Mechanics* 55 (1988) 98–103.

[15] J.F. Yau and S.S. Wang, An analysis of interface cracks between dissimilar isotropic materials using conservation integrals in elasticity, *Engineering Fracture Mechanics* 20 (1984) 423–432.

[16]   J.R. Rice, A path independent integral and the approximate analysis of strain concentration by notches and cracks, *Journal of Applied Mechanics* 35 (1968) 379–386.

[17]   X. Zhang, Y. Wang, J.-H. Im, and P.S. Ho, Chip-package interaction and reliability improvement by structure optimization for ultralow-*k* interconnects in flip-chip packages, *IEEE Transactions on Device and Materials Reliability* 12 (2012) 462–469.

[18]   D.T. Read, Y.W. Cheng, and R. Geiss, Morphology, microstructure, and mechanical properties of a copper electrodeposit, *Microelectronic Engineering* 75 (2004) 63–70.

[19]   A.A. Volinsky, J.B. Vella, and W.W. Gerberich, Fracture toughness, adhesion and mechanical properties of low-*k* dielectric thin films measured by nanoindentation, *Thin Solid Films* 429 (2003) 201–210.

[20]   P. Leduc, M. Savoye, S. Maitrejean, D. Scevola, V. Jousseaume, and G. Passemard, Understanding CMP-induced delamination in ultra low-*k*/Cu integration, in: *Proceedings of the IEEE 2005 International Interconnect Technology Conference, 2005*, IEEE, Burlingame, CA, 2005: pp. 209–211.

[21]   M.-Y. Tsai, P.-S. Huang, C.-Y. Huang, *et al.*, Investigation on Cu TSV-induced KOZ in silicon chips: Simulations and experiments, *IEEE Transactions on Electron Devices* 60 (2013) 2331–2337.

[22]   F.X. Che, W.N. Putra, A. Heryanto, A. Trigg, X. Zhang, and C.L. Gan, Study on Cu protrusion of through-silicon via, *IEEE Transactions on Components, Packaging and Manufacturing Technology* 3 (2013) 732–739.

Chapter 8

Mechanical response of solder joints under drop impact loads

8.1 Introduction

The increasing complexity of electronic products, coupled with the growing demand for portability and integration, has placed substantial pressure on the design and functionality of electronic packaging structures. Modern electronic packaging not only serves as a protective barrier for the delicate internal components but also plays a critical role in maintaining the thermal, electrical, and mechanical stability of the product during operation. As electronic devices become smaller and more powerful, the requirements for packaging materials and design strategies have become more stringent. Packaging must not only meet the functional needs of protection and integration but also ensure that devices can operate reliably in a variety of environmental conditions, from high-temperature environments to exposure to mechanical shocks and vibrations.

One of the most significant challenges faced by electronic packaging today is impact resistance, particularly during everyday use and transportation. Drop impacts are one of the most common real-world stressors that electronic products face. In fact, devices such as smartphones, laptops, and wearables are particularly vulnerable to mechanical stress resulting from accidental drops, which can cause severe damage to both the internal and external components. Such impacts can lead to cracked screens, broken circuits, solder joint failures, or even catastrophic damage to the integrated circuits themselves. These failures can disrupt the normal operation of the device, leading to functional defects, data loss, and, in some cases, total device failure, significantly affecting the user experience and reliability of the product [1–10].

The economic consequences of poor impact resistance can be considerable for both manufacturers and consumers. Manufacturers of high-end portable electronic products often face significant financial losses due to warranty claims and product returns caused by drop-induced failures. In many cases, damage resulting from drops is not covered under warranty, meaning that the consumer bears the full cost of repairs or replacement. This can result in substantial out-of-pocket expenses for the end user. Furthermore, the indirect costs of drop-induced failures—such as loss of vital data, downtime, and productivity losses—can be even more significant than the direct repair costs, particularly in business-critical devices. For instance, damage to a laptop can lead to the loss of crucial work, while the failure of a smartphone could result in missed business communications or other disruptions.

Given the increasing reliance on portable electronic devices for both personal and professional use, improving the impact resistance of electronic packaging has become a priority. To achieve this, it is essential to delve into the material properties, structural design, and failure mechanisms of packaging materials under impact loading conditions. Research on the behavior of packaging structures under various types of impact loading—such as free fall, edge impact, or corner impact—can provide valuable insights into the ways in which packaging can be optimized for durability. This includes the selection of more robust materials, the use of innovative shock-absorbing structures, and the design of packaging that can better dissipate energy during an impact event. Additionally, advanced simulation techniques, including finite-element modeling and experimental testing, are critical in assessing the performance of packaging designs and predicting failure modes under real-world impact conditions. By combining these approaches, it is possible to design more resilient packaging solutions that not only protect the internal components of the device but also enhance its overall reliability and longevity in the face of everyday mechanical stresses. Ultimately, such improvements will result in higher customer satisfaction, reduced repair costs, and a more favorable market position for manufacturers.

8.2 Test methodology and apparatus for drop impact testing

Drop impact testing includes both product-level and printed circuit board (PCB)-level tests. However, due to the poor repeatability of product-level tests, many researchers prefer PCB-level tests. For this purpose, specially designed drop impact testing machines are required. The Joint Electron Device Engineering Council (JEDEC) has recommended a drop impact testing machine in its standard draft JESD22-B111 [11], with its main structure shown in Figure 8.1.

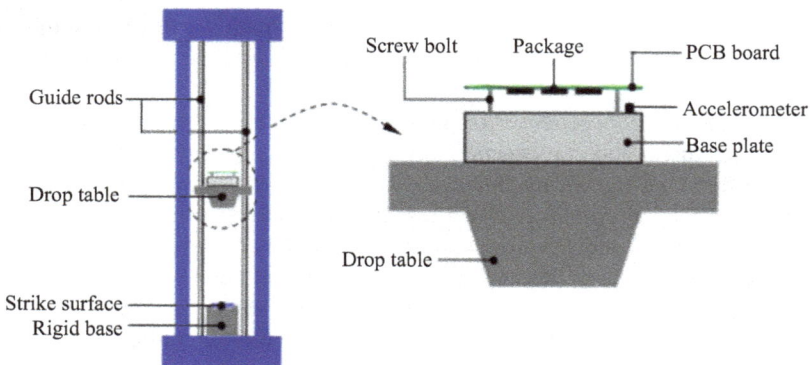

Figure 8.1   Test setup of board-level drop/impact

The testing machine primarily consists of guide rods, a drop table, and a rigid base. The rigid base is covered with materials such as felt to control the shape of the impact acceleration curve. The tested circuit board is fixed to the base plate on the upper part of the drop table using four screw bolts, with a 10 mm gap between the base plate and the circuit board to allow sufficient space for the circuit board to bend and deform. The test circuit board measures 100 mm × 48 mm × 1.6 mm and can accommodate up to 15 (3 × 5) packages. Figure 8.2 illustrates the numbering and positions of the packages. During the test, the drop table is released from a height of 1.5 m along the guide rods without initial velocity and falls onto the rigid base. This generates an acceleration pulse of a specified shape, amplitude, and duration that is applied to the drop table and transmitted to the tested PCB and packages through the base plate and fixing bolts. An accelerometer connected to the base plate records the peak value, duration, and shape of the acceleration curve in real time during the impact process.

JESD22-B110 provides testing standards for PCB-level impact acceleration pulses in portable products, as listed in Table 8.1 [12]. The standard recommends a half-sine acceleration pulse with a peak acceleration of 1500g and a duration of

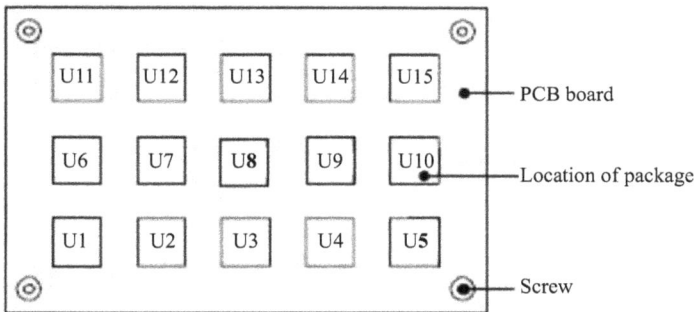

Figure 8.2   Layout of packages on PCB board

Table 8.1   Test conditions suggested by JEDEC

JEDEC shock pulse conditions	Peak accelerate G_m	Pulse duration T (ms)
H	2900	0.3
G	2000	0.4
B	1500	0.5
F	900	0.7
A	500	1.0
E	340	1.2
D	200	1.5
C	100	2.0

0.5 ms, corresponding to condition B in Table 8.1. Researchers or testers can also choose other conditions according to their needs.

8.3   Impact mechanics model and the Input-G method

The mechanical model of the drop impact process for board-level packaging can be illustrated using a dual spring–mass system [13], as shown in Figure 8.3. Here, m and k represent the mass and stiffness coefficient of the PCB, respectively, while M denotes the mass of the base plate and K indicates the stiffness coefficient of the impact surface on the rigid base. The motion equation of the system is given by

$$\begin{cases} M\ddot{Y} + (k + K)Y - ky = 0 \\ m\ddot{y} - kY + ky = 0 \end{cases} \tag{8.1}$$

The characteristic equation of the system is given by

$$\omega^4 - \frac{m(k + K) + Mk}{mM}\omega^2 + \frac{kK}{mM} = 0 \tag{8.2}$$

The eigenvalues can be derived from (8.2) as follows:

$$\frac{\omega^2}{\Omega^2} = \frac{1}{2}\left\{ \frac{Mk + m(K + k)}{mM} \pm \sqrt{\left[\frac{Mk + m(K + k)}{mM}\right]^2 - 4\frac{(K + k)k - k^2}{Mm}} \right\} \tag{8.3}$$

Given $K \geq k$, the solution can be obtained from (8.3) as follows:

$$\omega^2 \approx \frac{k}{m}\Omega^2 = \frac{K}{M} \tag{8.4}$$

The initial conditions are $y(0) = Y(0) = 0$, $\dot{y}(0) = \dot{Y}(0) = -V_0$. The solution to the motion equation is

$$y(t) = \frac{-V_0}{\omega(R_\omega^2 - 1)}\sin \omega t + \frac{-V_0 R_\omega^2}{\Omega(R_\omega^2 - 1)}\sin \Omega t$$

$$Y(t) = \frac{V_0}{\Omega}\sin \Omega t \tag{8.5}$$

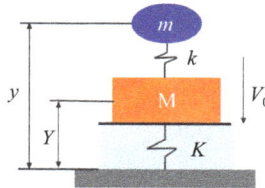

Figure 8.3   Two spring–mass systems

where $R_\omega = \omega/\Omega$, $V_0 = \sqrt{2gH}$, V_0 represents the velocity of the package drop impact onto the base, and H is the drop height. The acceleration of the substrate can be derived by taking the second derivative of (8.5)

$$G(t) = V_0 \sin \Omega t \tag{8.6}$$

The acceleration is a half-sine curve, as shown in Figure 8.4, with amplitude $G_m = V_0\Omega$ and period is $T = \frac{\pi}{\Omega}$. Equation (8.7) can be obtained according to (8.3)

$$G_m\sqrt{2gH\frac{K}{M}}T = \pi\sqrt{\frac{M}{K}} \tag{8.7}$$

Given that the stiffness of the substrate is significantly higher than that of the PCB, the half-sine acceleration transmitted from the substrate through bolts is effectively transferred to the PCB without significant loss. Therefore, during the numerical simulation of the package drop impact process, it is sufficient to establish a circuit board-level packaging model and apply the half-sine acceleration at the bolt locations on the PCB for computation. This approach is the underlying principle of the Input-G method.

The Input-G method involves applying the shock acceleration curve $G(t)$ as a load in the vertical direction at the four bolt locations on the PCB. Figure 8.5 illustrates the principle of this method. The $G(t)$ curve can be obtained either

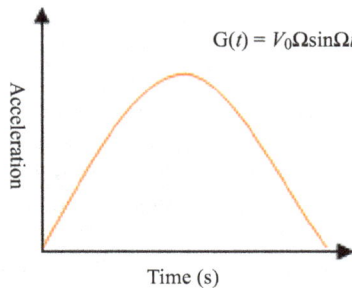

Figure 8.4 Half of the sine acceleration curve

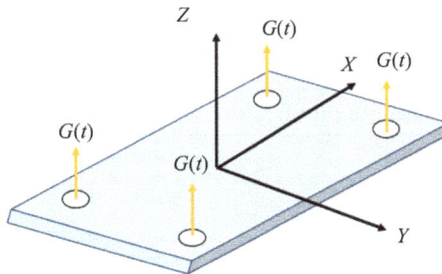

Figure 8.5 Input-G method

through measurements from a testing machine or selected from the testing standards recommended by JEDEC, as per the need. In this chapter, the acceleration is chosen according to the JEDEC standard JESD22-B110 under condition *H*, as detailed in Table 8.1. Since $G(t)$ is derived from actual experimental data, this approach circumvents technical challenges associated with full-scale simulation (such as adjusting parameters of contact surfaces and defining contact types), thereby enhancing the comparability of numerical simulation results and reducing computational time. Furthermore, when employing the Input-G method, it is sufficient to construct a packaging model, allowing for the establishment of a more comprehensive board-level packaging model. This facilitates a detailed analysis of the entire dynamic response process of drop impact on fine packaging components, such as solder joints.

8.4   Numerical simulation study on drop impact of solder joints

8.4.1   Establishment of the structural model and parameters

During the analysis of stress on board-level packaging under drop impact conditions, the assembly board in the model is conceptualized as a three-layer structure composed of the upper substrate, a solder ball array, and the PCB. The solder ball array employs a 6 × 6 configuration with a spacing of 0.75 mm. This model considers various details including the PCB, copper (Cu) pads, solder balls, substrate, chip, and encapsulation resin (MC). The design of the Cu pads is surface-mount device (SMD) on the substrate side and non-solder mask defined (NSMD) on the PCB side. The geometric dimensions of the model are presented in Table 8.2. The modeling process adheres to the principle of progression from small to large and from the inside out; thus, solder joints are modeled first, followed by the Cu pads, SMD, and NSMD. The upper Cu pad is embedded within the solder joint, while the lower Cu pad is integrated into the SMD, achieved through the Overlap command in Boolean operations to ultimately create the PCB, chip, and encapsulation resin. To obtain a relatively regular mesh, this study employs the sweep mesh technique with custom element lengths. Given that the stress conditions of the solder joints are the focus of this research, a finer mesh is applied at the solder joint locations.

Table 8.2   Dimension of package model

Materials	Length (mm)	Width (mm)	Height (mm)
PCB	100	48	1.6
Substrate	7	7	0.2
SMD, NSMD	7	7	0.04
Die	4	4	0.26
MC	7	7	0.75
Solder	ϕ0.35		0.15

The PCB, being comparatively larger relative to the solder joints, utilizes a coarser mesh in regions farther away from the solder joints. The established three-dimensional finite-element model is illustrated in Figure 8.6.

In this chapter, the SOLID164 element, which possesses explicit analysis capabilities, is chosen for meshing the solid model. The SOLID164 element is an eight-node hexahedral element that can degenerate into six-node wedge elements or five-node tetrahedral elements by repeating certain nodes. There are two algorithms for solid elements: (1) KEYOPT(1)=1, which utilizes single-point integration and hourglass control, providing savings in computational time while enhancing reliability under conditions of significant deformation; (2) KEYOPT(1) = 2, which employs a $2 \times 2 \times 2$ multi-point Gauss integration and does not produce zero energy modes, thereby eliminating the need for hourglass control. For certain materials, such as foam materials, the computational results may be improved.

For the explicit dynamic analysis of the drop impact problem in packaging, single-point Gaussian integration is utilized for element calculations to enhance computational efficiency. However, single-point integration elements are prone to generating zero-energy modes, known as hourglasses (which are zero-energy deformation modes exhibiting frequencies that are significantly higher than the global structural response). If the hourglass modes are not controlled during dynamic response calculations, numerical oscillations may occur. To mitigate zero-energy modes, this study employs hourglass viscous damping in LS-DYNA. The parameter values of the materials used in the model are summarized in Table 8.3, and all material models are linear elastic.

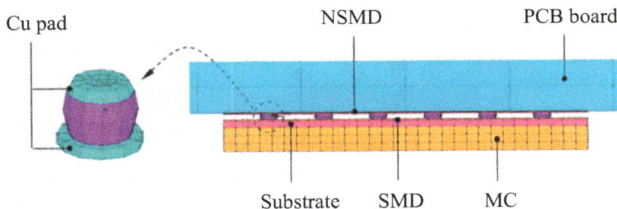

Figure 8.6   Local finite-element model of the BGA package and PCB board

Table 8.3   Material properties

Materials	Modulus (MPa)	Poisson's ratio	Density (10^{-3} g/mm^3)
Die	13,100	0.3	2.33
Mold compound	25,506	0.3	1.97
Cu pad	117,000	0.3	8.94
PCB	16,850	0.11	1.82
Solder	34,000	0.363	8.41
Solder mask	5000	0.3	1.15
Substrate	3499	0.3	3.97

8.4.2   Boundary conditions

In the Input-G method, the shock load imposed on the PCB is applied in the form of acceleration at the vertical direction of the four bolt locations on the PCB. Consequently, in the finite-element model, the acceleration load can be applied to the vertical degree of freedom of the nodes within the bolt clamp area. The clamping effect of the bolts on the PCB influences its deformation; hence, to account for this effect, appropriate displacement constraints should be applied in the horizontal direction at the nodes within the model's bolt clamp area. To investigate the influence of different constraints on the results, this chapter considers three constraint scenarios: the first is no constraints applied, the second is horizontal displacement constraints imposed on the upper surface of the PCB within the four bolt clamp areas, and the third is horizontal displacement constraints applied on both the upper and lower surfaces of the clamp areas.

Figure 8.7 presents the delamination stress response curves of the stress-maximized solder joints under the three conditions. It is evident that under condition (a), where no constraints are applied, the stress level of the solder joints is significantly higher and does not exhibit notable attenuation over time. Under condition (b), the stress value of the solder joints decreases by approximately 70 MPa, showing significant attenuation but with more pronounced fluctuations. Under condition (c), the maximum stress is the same as in condition (b), with significant attenuation and a smoother curve that aligns more closely with the

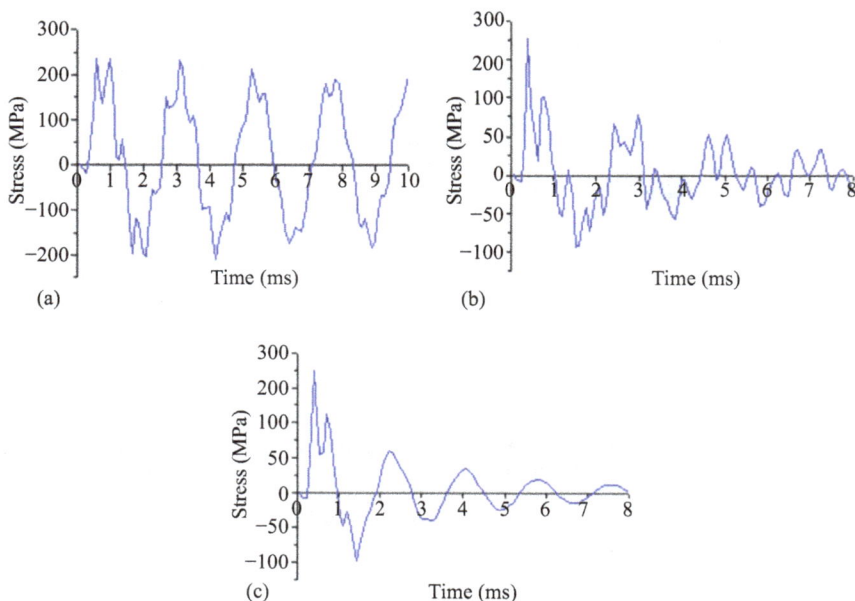

Figure 8.7   Stress responses under different bolt effects

experimental results. Therefore, adopting the third displacement constraint condition is deemed more reasonable.

8.4.3   *Results and discussion*

The solver used for the simulation process is LS-DYNA. Figure 8.8 presents the overall stress distribution of the solder joints along with an enlarged view of the maximum stress solder joint. It is observed that the solder joint at the outer corner experiences the greatest stress. Therefore, the stress conditions at the outer corner solder joints should be regarded as a critical area of concern.

Figure 8.12 illustrates the dynamic response curve of the maximum tensile stress node (node 959) located at the far corner of the solder joint, as shown in Figure 8.9. The stress components presented include the axial stress (S_x) along the length of the PCB, the normal stress in the Z direction (S_z), the shear stress (S_{xz}), the first principal stress (S_1), and the von Mises stress (S_{eqv}). These stress components exhibit periodic variation in response to the vibrations of the PCB, with the magnitude of the peeling stress being comparable in both positive and negative

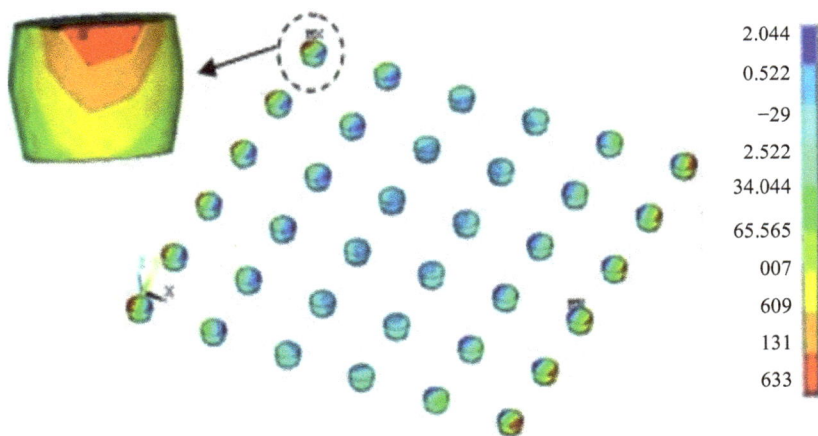

2.044
0.522
−29
2.522
34.044
65.565
007
609
131
633

Figure 8.8   Stress in solder joints

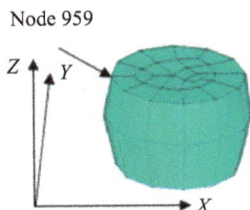

Node 959

Figure 8.9   Outermost solder joint

directions. The curve for the first principal stress exhibits a similar shape to that of the peeling stress, while the von Mises stress also reflects the development trend of the peeling stress. The values of S_x and S_1 are relatively low. Consequently, the peeling stress S_z is selected as the primary focus for analyzing the stress conditions of the solder joints. As observed in Figure 8.7c, the solder joints initially experience tensile stress, reaching its peak at 0.4 ms, before transitioning into compressive stress and alternating thereafter, eventually exhibiting a gradual attenuation. The maximum stress occurs at the junction of the far corner solder joint and the Cu pad on one side of the PCB, reaching a value of 192 MPa (Figure 8.10).

Figure 8.11 demonstrates the variation of deflection at the center of the PCB during the impact process. It is evident that the PCB undergoes upward and downward bending deformation during the impact, with a rapid attenuation over time. As observed in Figure 8.11, the peaks of solder joint stress and PCB deflection occur simultaneously, indicating that during drop impacts, the stress in the solder joints is closely related to the deformation of the PCB. Many researchers believe that the stress in the solder joints can be effectively explained by the mechanism model presented in Figure 8.12. Due to the differences in bending

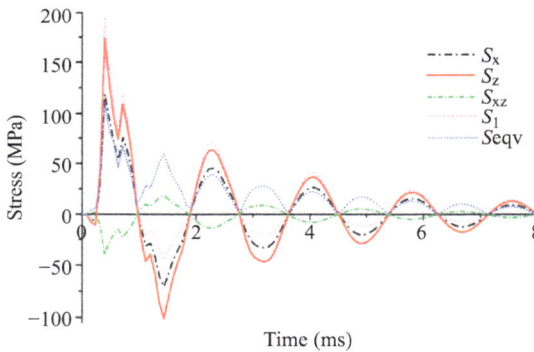

Figure 8.10   Dynamic response of solder joint

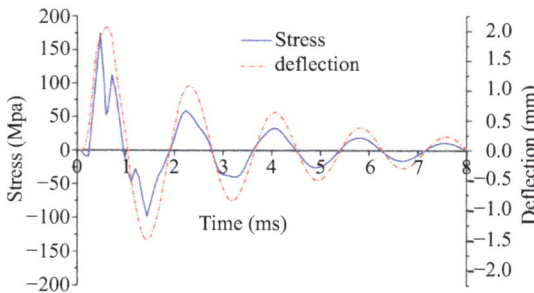

Figure 8.11   Comparison between stress in solder joints and deflection of PCB

Figure 8.12   *Deflection causes (a) tension stress in solder joints by PCB downward bending and (b) compressive stress by PCB upward bending*

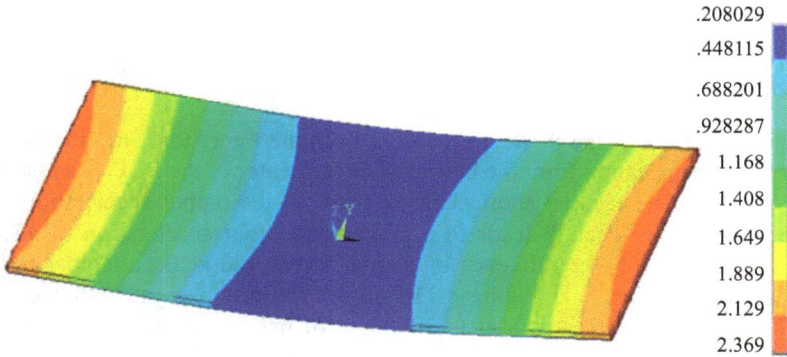

.208029
.448115
.688201
.928287
1.168
1.408
1.649
1.889
2.129
2.369

Figure 8.13   Deflection of PCB

stiffness between the PCB and ball grid array (BGA) packages, this bending deformation of the PCB induces peeling stresses within the solder joints. When the PCB bends downwards, tensile stresses are generated in the solder joints, with greater tensile stress observed in the joints on both sides. Conversely, when the PCB bends upwards, corresponding solder joints experience compressive stresses. Throughout the impact process, the repeated bending of the PCB results in alternating tensile and compressive stresses in the solder joints, as illustrated in Figure 8.11.

This mechanism model effectively explains why the maximum peeling stress occurs at the solder joint located at the far corner of the package. Figure 8.13 presents a deformation cloud image of the PCB, revealing that the PCB experiences bending deformation in both the longitudinal and transverse directions. The combined effects of these two forms of deformation result in the maximum peeling stress at the corner solder joints. Furthermore, due to the significant deformation in the longitudinal direction of the PCB, the stress levels in the outermost rows of solder joints in that direction are also elevated.

The stress in the solder joints is closely related to the deformation of the PCB. Furthermore, the output acceleration serves as a significant parameter in describing the dynamic response of the PCB. Figure 8.14 presents the output acceleration

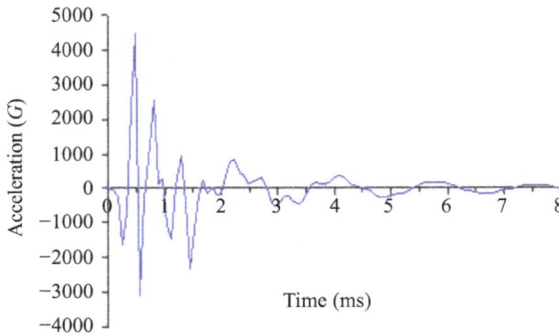

Figure 8.14   Out acceleration at the center of PCB

curve at the center of the PCB. It is evident from the figure that after the drop, the stress wave propagating from the bolt reaches the center of the PCB in a very short time, resulting in an initial short negative pulse in the output acceleration at the center, which subsequently leads to a maximum output acceleration of 4600 G. Following this, under the influence of inertial forces, the PCB undergoes bending deformation. As shown in Figure 8.11, the maximum stress in the solder joints occurs at 0.40 ms, rather than at the moment of maximum input acceleration at 0.15 ms. This disparity is attributed to the substantial bending deformation of the PCB caused by inertial forces after the drop impact, which in turn leads to the maximum solder joint stress. Inertial forces play a crucial role in the failure of solder joints. From the above analysis, it is clear that the difference in bending stiffness between the PCB and the BGA package, along with the inertial forces from the package, significantly influences the stress in the solder joints.

8.5   Chapter summary

The dynamic response of solder joints in board-level BGA packages during drop impacts was analyzed using the explicit finite-element dynamic analysis method. The study highlights several key observations that provide insights into the mechanical behavior of solder joints and the PCB under high-impact conditions, which are critical for evaluating the reliability and durability of BGA packages. First, the treatment of constraint conditions at the fixed bolt locations on the PCB is crucial for accurate results, particularly when using the Input-G method for calculating dynamic responses. Inadequate modeling of these constraints can lead to significant errors in the prediction of stress distribution and deformation patterns. The study found that applying horizontal displacement constraints to both the upper and lower surfaces of the PCB in the regions surrounding the four fixed bolts provides a more reasonable and realistic boundary condition. This approach better simulates the true behavior of the PCB during drop impacts, as the bolts serve to limit the movement of the PCB while allowing it to deform under impact.

Second, the analysis revealed that the maximum stress in the solder joints occurs immediately after the initial impact. The highest stresses observed were peeling stresses, which are most significant at the junction between the corner solder joint and the Cu pad on one side of the PCB. This observation is important as the corner solder joints are typically more vulnerable to failure due to the concentrated stresses at the interface between the solder ball and the Cu pad. The local stress concentrations in these regions are exacerbated by the PCB's deformation, which results in a higher likelihood of solder joint cracking or detachment in the corners during drop events.

In addition, the study identified that the stress levels in the outermost rows of solder joints are elevated, particularly along the longitudinal direction of the PCB. This is primarily due to the significant bending and deformation of the PCB during the drop impact, which creates a differential strain distribution across the board. As the PCB bends, it induces higher stresses in the solder joints at the outermost rows, making them more susceptible to failure. The research also highlights that the peak solder joint stress and the peak PCB deformation occur simultaneously, indicating that the bending deformation of the PCB plays a critical role in generating and amplifying the stress in the solder joints. This correlation underscores the importance of considering the PCB's global deformation behavior when evaluating solder joint reliability under dynamic loading conditions.

The study further explored the relationship between the PCB's bending direction and the stress state in the solder joints. When the PCB bends downward (toward the component side), tensile stresses are generated in the solder joints. This effect is particularly pronounced in the solder joints located on both sides of the PCB, where the tensile stresses are most concentrated. On the other hand, when the PCB bends upward (away from the component side), the solder joints experience compressive stresses. The tensile and compressive stresses generated during bending significantly influence the mechanical integrity of the solder joints, especially in terms of the risk of solder joint cracking, peeling, or other forms of mechanical failure.

Overall, the results from this analysis provide valuable insights into the stress distribution and deformation behavior of solder joints in BGA packages under drop impact conditions. The findings suggest that the most vulnerable solder joints are those located at the corners and the outermost rows, with the bending of the PCB being a major contributing factor to stress concentration. Understanding these dynamics is essential for improving the design and reliability of BGA packages, particularly in applications subject to frequent mechanical shocks or drop events. Additionally, the study emphasizes the need for careful consideration of boundary conditions and PCB deformation in simulations to accurately predict the performance of solder joints under dynamic loads.

References

[1] Y.H. Yau and S.N. Hua, A comprehensive review of drop impact modeling on portable electronic devices, *Applied Mechanics Reviews* 64 (2011) 020803.

[2] B. Zhang, J.S. Xi, P.K. Liu, and H. Ding, Failure analysis of board-level Sn-Ag-Cu solder interconnections under JEDEC standard drop test, *Journal of Electronic Materials* 42 (2013) 2848–2855.

[3] N. Muthuram and S. Saravanan, Free fall drop impact analysis of board level electronic packages, *Microelectronics Journal* 129 (2022) 105601.

[4] M. Mao, W. Wang, C. Lu, F. Jia, and X. Long, Machine learning for board-level drop response of BGA packaging structure, *Microelectronics Reliability* 134 (2022) 114553.

[5] J. Zhao and L.J. Garner, Mechanical modeling and analysis of board level drop test of electronic package, in: *56th Electronic Components and Technology Conference 2006*, IEEE, San Diego, CA, USA, 2006: pp. 436–442.

[6] E.H. Wong and Y.-W. Mai, New insights into board level drop impact, *Microelectronics Reliability* 46 (2006) 930–938.

[7] X. Long, Y. Hu, T. Su, and C. Chang, Numerical simulation of impact response of board-level packaging structure, in: S. Li (ed.), *Computational and Experimental Simulations in Engineering*, Springer International Publishing, Cham, 2024: pp. 1443–1453.

[8] C.-L. Yeh, T.-Y. Tsai, and Y.-S. Lai, Transient analysis of drop responses of board-level electronic packages using response spectra incorporated with modal superposition, *Microelectronics Reliability* 47 (2007) 2188–2196.

[9] T.-Y. Tsai, C.-L. Yeh, Y.-S. Lai, and R.-S. Chen, Transient submodeling analysis for board-level drop tests of electronic packages, *IEEE Transactions on Electronics Packaging Manufacturing* 30 (2007) 54–62.

[10] S.-S. Yeh, P.Y. Lin, M.C. Yew, *et al.*, Ultra-thin package board level drop impact modeling and validation, in: *2019 IEEE 69th Electronic Components and Technology Conference (ECTC)*, 2019: pp. 1550–1555.

[11] L.B. Tan, S.K.W. Seah, E.H. Wong, X. Zhang, V.B.C. Tan, and C.T. Lim, Board level solder joint failures by static and dynamic loads, in: *Proceedings of the 5th Electronics Packaging Technology Conference (EPTC 2003)*, IEEE, Singapore, 2003: pp. 244–251.

[12] Y.C. Ong, V.P.W. Shim, T.C. Chai, and C.T. Lim, Comparison of mechanical response of PCBs subjected to product-level and board-level drop impact tests, in: *Proceedings of the 5th Electronics Packaging Technology Conference (EPTC 2003)*, IEEE, Singapore, 2003: pp. 223–227.

[13] E.H. Wong, Y.-W. Mai, and S.K.W. Seah, Board level drop impact—Fundamental and parametric analysis, *Journal of Electronic Packaging* 127 (2005) 496–502.

Index